Internationale Reihe Agribusiness

Band 26

Hrsg. von Ludwig Theuvsen

ISSN 1869-9316

The German Dairy Sector: Internationalization – Competitiveness – Supply Chains

Dissertation
to obtain the Doctor of Agriculture (Ph. D.) degree
in the Ph. D. Program for Agricultural Sciences in Göttingen (PAG)
at the Faculty for Agricultural Sciences,
Georg-August-Universität Göttingen

presented by

Johannes Meyer

born on 14.04.1988 in Nienburg/Weser

Göttingen, October 2019

Bibliografische Information der Deutschen Nationalbibliothek
Die Deutsche Nationalbibliothek verzeichnet diese Publikation in der Deutschen Nationalbibliografie; detaillierte bibliographische Daten sind im Internet über http://dnb.d-nb.de abrufbar.
1. Aufl. - Göttingen: Cuvillier, 2020
Zugl.: Göttingen, Univ., Diss., 2019

D7

1. Supervisor: Prof. Dr. Ludwig Theuvsen
2. Co-Supervisor: Prof. Dr. Jan-Henning Feil
3. Examiner: Prof. Dr. Achim Spiller

Date of disputation: 4th of November 2019

Nonnenstieg 8, 37075 Göttingen
Telefon: 0551-54724-0
Telefax: 0551-54724-21
www.cuvillier.de

1. Auflage, 2020
Gedruckt auf umweltfreundlichem, säurefreiem Papier aus nachhaltiger Forstwirtschaft.

ISBN 978-3-7369-7212-4
eISBN 978-3-7369-6212-5

Geleitwort

Die Milchwirtschaft ist gemessen am Umsatz nach der Fleischwirtschaft der zweitwichtigste Teilbereich des deutschen Agribusiness. Die für Landwirte oft unbefriedigenden Auszahlungspreise für Milch hängen überwiegend von der Angebots- und der Nachfragesituation auf dem Weltmarkt ab. Speziell im Standardproduktebereich werden die Preise maßgeblich durch die internationalen Kostenführer, allen voran Neuseeland, mitbestimmt. Für ein Land wie Deutschland mit einem Selbstversorgungsgrad bei Milch von mehr als 110 %, das unter Berücksichtigung der Einfuhren von Milchprodukten letztlich rund 50 % der im Inland erzeugten Milch auf ausländischen Märkten absetzen muss, ist die Frage der internationalen Wettbewerbsfähigkeit von herausragender Bedeutung. Von ihrer Beantwortung hängt es ab, ob deutsche Milcherzeugnisse auch zukünftig erfolgreich, d.h. profitabel und in ausreichender Menge, im Ausland abgesetzt werden können. Die internationale Wettbewerbsfähigkeit des gesamten Sektors ist zudem eine wesentliche Determinante der Zukunftsfähigkeit und der Einkommenschancen jedes einzelnen Milcherzeugers in Deutschland und damit der Perspektiven vieler ländlicher Räume.

Herr Johannes Meyer hat die beschriebene Situation zum Anlass genommen, sich näher mit der Frage zu befassen, wie sich vor dem Hintergrund veränderter Rahmenbedingungen, etwas des zwischenzeitlichen Wegfalls der Milchquote, die internationale Wettbewerbsfähigkeit der deutschen Milchwirtschaft insgesamt sowie in ausgewählten Produktgruppen im Zeitablauf entwickelt hat und wie sich die unterschiedlichen Internationalisierungsstrategien von Molkereien auf den jeweiligen Unternehmenserfolg und damit auch auf die Fähigkeit der Unternehmen, aus Sicht der Landwirtschaft angemessene Auszahlungspreise zu gewährleisten, auswirken. Ergänzend ist Herr Meyer der Frage nachgegangen, welche Faktoren Einfluss auf die strukturelle Entwicklung der Milcherzeugung nehmen. Schließlich hat er auch exemplarisch Fragen der Supply Chain-Integration am Beispiel des brasilianischen und des deutschen Milchsektors aufgegriffen.

Herr Meyer hat sich einer aus Sicht der gesamten Wertschöpfungskette ‚Milch' in hohem Maße praktisch bedeutsamen und wissenschaftlich reizvollen Problemstellung zugewandt. Die von ihm aufgegriffene Forschungsfrage ist zudem nur vergleichsweise selten empirisch untersucht worden. Seine Dissertationsschrift vermittelt daher neue, zum Teil überraschende Einblicke in einen bedeutsamen Bereich der agrarökonomischen Forschung. Aus diesem Grund wünsche ich der vorliegenden Arbeit die verdiente Aufmerksamkeit in der milchwirtschaftlichen Praxis wie auch der Scientific Community.

Göttingen, im Mai 2020 Prof. Dr. Ludwig Theuvsen

Danksagung

Nach dreieinhalb Jahren ist die Dissertation fertiggestellt und alle damit verbundenen Prüfungen und Voraussetzungen erfüllt. Sie ist während meiner Tätigkeit als wissenschaftlicher Mitarbeiter am Department für Agrarökonomie und rurale Entwicklung der Georg-August-Universität Göttingen entstanden. An dieser Stelle möchte ich mich ganz herzlich bei allen Personen bedanken, die mich auf unterschiedlichste Art und Weise unterstützt haben und zum erfolgreichen Abschluss der Promotion beigetragen haben.

Zu allererst möchte ich mich bei meinem Doktorvater Herrn Prof. Dr. Ludwig Theuvsen bedanken, der mir die Möglichkeit gegeben hat, in seinem Team mitzuarbeiten und meine Doktorarbeit bei ihm zu schreiben. Vielen Dank für das entgegengebrachte Vertrauen, die guten Ratschläge, die Freiheiten dieses spannende Themenfeld nach meinen Vorstellungen zu bearbeiten und für ein stets offenes Ohr. Es war eine sehr schöne Zeit am Lehrstuhl an dem ich mich immer sehr wohl gefühlt habe und auf die ich gerne zurückblicken werde. Mein weiterer Dank gilt Herrn Prof. Dr. Jan-Henning Feil für die gute Zusammenarbeit und guten Ratschläge und dem damit ebenfalls maßgeblichen Anteil an dem Gelingen dieser Arbeit. Ebenfalls vielen Dank an Dr. Christian Schaper für die gute Betreuung und Zusammenarbeit. Herrn Prof. Dr. Achim Spiller möchte ich meinen herzlichen Dank aussprechen, dass er als Drittprüfer am Tag der Disputation das Prüfungskomitee vervollständigt hat. Ebenfalls danken möchte ich Dr. Verena Otter, die mir in wichtigen Situationen den nötigen Anreiz zum vorwärtskommen gegeben hat.

Ein ganz herzlicher Dank gilt auch meinen Kollegen und Freunden aus der 11. und 10. Etage, insbesondere Caetano, Caro, Mira, Johannes, Wienand, Dirk, Louisa, Luis, Gloriana und Thuy, sowie Andy aus der Agrartechnik für die tolle Zeit, konstruktive und unterhaltsame Gespräche und Mittagspausen und stets einem offenen Ohr.

Bedanken möchte ich mich auch bei Herrn Prof. Dr. Sebastian Lakner für die gute Unterstützung im Doktorandenseminar und die weitere Zusammenarbeit. Ebenfalls bedanken möchte ich mich bei Colin Gordon, der den Sprachcheck für die gesamte Dissertation gemacht hat und während der gesamten Zeit immer auch kurzfristig erreichbar war.

Ganz herzlich bedanken möchte ich auch unserem langjährigen Mitarbeiter und Freund Mirek, der zusammen mit meinen Eltern den landwirtschaftlichen Betrieb am Laufen gehalten hat und ohne dessen Unterstützung meine Abwesenheit für die Doktorarbeit so nicht möglich gewesen wäre.

Zu guter Letzt möchte ich mich ganz besonders bei meiner Familie und hier insbesondere bei meinen Eltern bedanken, die mich bei all meinen Entscheidungen stets unterstützt haben, stets für mich da waren wenn es nötig war und in meiner Abwesenheit den Betrieb am Laufen gehalten haben. Vielen Dank, dass ihr es mir ermöglicht habt diese Promotion umzusetzen und den landwirtschaftlichen Betrieb fortzuführen.

Vielen Dank Euch allen für die wundervolle Zeit in Göttingen, auf die ich immer sehr gerne zurückblicken werde.

Göttingen im Mai 2020 Johannes Meyer

Table of contents

I Introduction

"Some are heading in the wrong direction."

With these words Peter Stahl, CEO of Hochland SE, one of the largest dairies in Germany and CEO of the Dairy Industry Association, described the orientation of German dairies towards the export of inadequately differentiated and interchangeable dairy products to countries with the necessary natural resources for milk production. These are, for example, countries in Western Africa, such as Nigeria and the Senegal (EMB, 2019; Cornall, 2019).The statement also refers to the fact that these exports are at least partly realized on the basis of feed imports and are associated with negative externalities in Germany, such as high nutrient surpluses. In case of milk production these are, for example, the regions around the North Sea coast in Lower Saxony (Chamber of Agriculture, 2019; top agrar, 2019).

Looking at the figures, the German dairy industry has developed quite positively in recent years. In 2018 turnover was € 28.1 billion, whilst exports amounted to € 9 billion. While sales are strongly influenced by product prices, the dairy industry has shown a positive trend over recent years in terms of both the number of companies and the number of employees (see Figure 1). In 2018, 244 companies assigned to it employed 44,252 people, whereas investments amounted to € 711 million in 2017 (Destatis, 2019a, 2019c). Investments remain at a high level despite a decline when compared to 2016. Due to the frequently rural situation of these companies, the dairy industry is of great economic importance as part of German agribusiness, especially for rural and, in part, structurally weak regions (Janze *et al.*, 2018, 2019).

Despite this development, the German dairy industry is under considerable competitive pressure. This is illustrated by the industry's average EBIT margin of around only 2%, which is well below the average for the food industry. One of the main reasons for this low profitability is the high price pressure exerted by the powerful food retail market. In 2015, the market share of the four largest food retailers in Germany was 85% (Schlippenbach and Pavel, 2011; Bundeskartellamt, 2015; Brokelmann and Gäde, 2017). High competitive pressure can also be seen in mergers and acquisitions, such as the merger of Humana and Nordmilch to form the new German market leader Deutsches Milchkontor (DMK) in 2011, or the takeover of the almost insolvent OMIRA

Molkerei by the French Lactalis Group in 2017 (Handelsblatt, 2011; dlz Agrar-maganzin, 2017).

Figure 1. Development of turnover, export and employees in the German dairy industry from 2008 to 2018

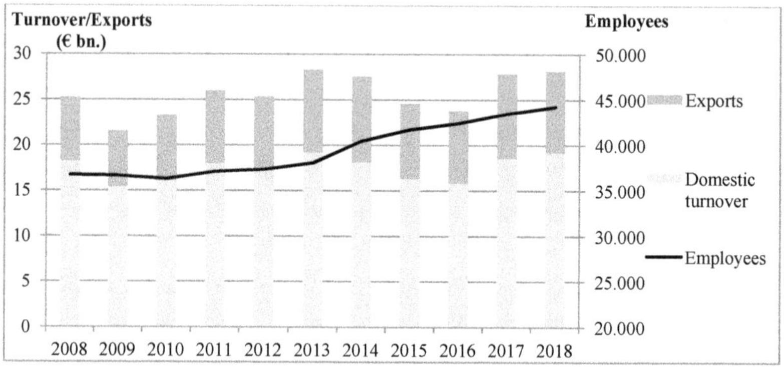

Source: Authors' depiction based on Destatis, 2019a; Destatis, 2019b, Destatis, 2019c.

The global dairy industry has experienced a strong wave of internationalization generated by growing demand (primarily from developing countries), policy drivers (such as the General Agreement on Tariffs and Trade (GATT) in the middle of the 1990's and the liberalization of the Common Agricultural policy (CAP)) and price drivers (for example in 2001 and 2007) (Guillouzo and Ruffio, 2005; Heyder *et al.*, 2011; Vitaliano, 2016). A similar trend applies to the German dairy industry; many dairies generate more than 40%, a few even more than 50%, of their total sales outside of Germany (Meyer and Theuvsen, 2017). As a whole, the German dairy industry had a total share of foreign sales of 31.8% in 2018. As a result, the export ratio rose from 27.8% in 2008 by four percentage points. However, it peaked in 2014 at 34.2%. Since then, the export ratio has declined (Destatis, 2019a; Destatis, 2019b). Even if the export ratio initially appears low compared to other sectors, such as the agricultural machinery industry (74.5% in 2017), the importance of export business for the German dairy industry becomes clearer when considering volumes (Janze *et al.*, 2019). According to the Dairy Industry Association, 49% of the raw milk processed in Germany was exported in the form of milk and milk products (MIV, 2019a). Internationalization has thus become the most important driver in the development of the dairy industry (Theuvsen *et al.*, 2010).

Internationalization is unlikely to lose any importance in the future. Although current trends such as; increasing quality orientation, sustainability, transparency and the increasing demand for organic food, offer companies the opportunity to position themselves in newly emerging niche markets with appropriate products, the whole process is taking place against the background of a saturated domestic market (Bratschi and Feldmann, 2005; Brümmer *et al.*, 2018; Mehlhose *et al.*, 2019). Thus, due to a self-sufficiency rate from 100% in butter to 163% in dried milk products, exports will remain a necessary outlet for the German dairy sector (BMEL 2019). Internationalization is also likely to remain a key driver in the future against the background of forecast increases in global demand and production, which will essentially take place in other parts of the world, in particular Asia and Africa (OECD and FAO, 2019). For the German dairy industry, this development can already be seen today from the list of most important export targets. The member states of the European Union continue to be the most important trading partners, accounting for 83.3% of total export sales in 2017. However, the importance of exports to third countries is increasing step by step. The share of non-EU exports rose from 12.8% in 2001 to 16.7% in 2017, although the share varies greatly depending on the product group under consideration (Trademap, 2019; MIV, 2019b).

Most dairy companies serve markets outside Germany with classic exports. This applies in particular to German dairy cooperatives, which process an estimated 60 to 70% of German milk production. Studies show that cooperative dairies, which specialize heavily in the export of undifferentiated dairy products, for instance milk powder or butter, have a particularly low added value (Jürgens, *et al.*, 2015; Hörl and Hess, 2017). Management theory proposes that an export strategy can only be successfully implemented if there are competitive advantages in the home country over competitors in other countries (Grant and Nippa, 2006). Undifferentiated products compete on the market through low prices (Porter, 1980).

With regard to competitive advantages in the dairy industry, it is important to note that due to the limited transportability and storability of raw milk, processing milk is to a large extent linked to domestic milk production (Friedrich, 2010). In the past, EU milk production was strongly regulated by agricultural policy measures, for instance by a quota regime. In the course of increasing liberalization of the Common Agricultural Policy (CAP), milk production and the dairy industry had to orient themselves more strongly towards international competition (Lassen, 2011; Tangermann and Cramon-Taubadel, 2013). With regard to the liberalization of the milk market, the biggest step was probably the abolition of the milk quota in 2015 (Hörl and Hess, 2017). However, the measures implemented in the run-up to the abolition, such as the expansion of milk quota volumes or the improved tradability of milk quotas via quota exchanges, had already led to a significant increase in milk production and to a regional shift in milk production in Germany and several other EU member states (Brümmer and Loy, 2004; Kirner, 2009; Betzholz, 2010; Eurostat, 2019). At the same time, the structural change in milk production continues to progress, while demands are constantly increasing, for example in the form of stricter environmental legislation as well as stricter requirements for husbandry conditions, on the part of consumers. (Lutter, 2009; Lindena *et al.*, 2018). As a consequence, the number of dairy farms in Germany has been decreasing and average herd size has been increasing over the past few decades (see Figure 2).

Figure 2. Development of number of farms and cows per farm from 1992 to 2018 in Germany

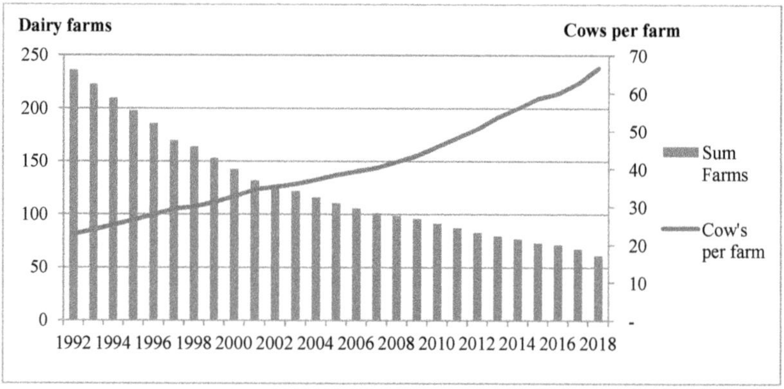

Source: Authors' depiction based on ZMB, 2012; Destatis, 2011–2019d.

The research objectives of this work are manifold. Due to the liberalization outlined above and the resulting new competitive environment, new framework conditions have emerged both for the German dairy industry itself and for its upstream milk production. Against this background, this dissertation examines which factors can influence the development of milk production after the end of the milk quota and how they affect it. With regard to the dairy industry itself, this dissertation examines how competitiveness has developed on the international market for milk and dairy products and how different internationalization strategies affect corporate success. Based on these results, strategic

lines of development for the German dairy industry are illustrated, as well as implementation methods based on a concrete example. These research objectives are addressed in a number of papers (Chapters II to VI).

Chapter II (*"Intensive Dairy Farming in Northern Germany: Development and Impact of the New Fertilizer Act"*, published in Agrarian Perspectives XXVI – Competitiveness of European Agriculture and Food Sectors – Proceedings of the 26th International Scientific Conference (2017)) examines the regional development of milk production in Germany and the effects of German legislation on farmers' nutrient management which was amended in March 2017. Data from the 2010 agricultural structure survey at the district level and secondary data on German milk production were used for the analysis.

Chapter III *("Drivers of Large Herd Sizes in German Dairy Farming: Development and Outlook"*, submitted to Outlook on Agriculture) analyzes factors influencing the development of large herds in German dairy farming. The influence of the milk quota, milk price, capital costs, prospects and opportunity costs on the development of large herds in the period from 1992 to 2018 is analyzed. The analysis is done quantitatively in the form of an OLS model with lagged variables, as well as qualitatively. Based on the results and other studies an assessment of the future development is given.

Chapter IV (*"Assessing the International Competitiveness of the German Dairy Industry by Analysing Foreign Trade"*, in preparation for submission to Agricultural Economics (AGRICECON) examines the international competitiveness of the German dairy industry compared to nine major competitors on the international dairy export markets. The analysis covers a period of 17 years from 2001 to 2017 and takes into account aggregated exports as well as six different groups of dairy products classified on the basis of the Harmonized System of the World Customs Organization. The article first defines the concept of international competitiveness. Although there are numerous publications on this subject, there is still no generally applicable definition of this term (Gries and Hentschel, 1994; Feurer and Chaharbaghi, 1994; Trabold, 1995; Baade, 2007). On the basis of the definition "the ability to sell their products on foreign markets for sustained, full cost covering prices, while gaining market share or at least maintaining its market share" based on Utzig (1987), Martin *et al.* (1991) and Weindlmaier (2000), we analyze international competitiveness using Relative Export Advantage (RXA), Relative Import Advantage (RMP) and Relative Trade Advantage (RTA)

(Vollrath, 1991). In addition, the export structure of the countries in question is analyzed over time and the most important trading partners are identified. On this basis, important developments in product group dependency and future developments are derived and the implications of the results for the German dairy industry are discussed.

Chapter V (*"Internationalization Strategies in the German Dairy Industry and their Influence on the Economic Performance of Firms"*, published in International Journal of Food System Dynamics 10 (4): 332–346 (2019)), the paper examines the internationalization strategies present in the German dairy industry and their influence on the economic success of companies. The analysis includes 18 companies. 16 of these are headquartered in Germany, 2 are from abroad but with their German subsidiaries have a significant share in milk processing in Germany. The study is based on the consolidated annual accounts of the companies in the period from 2010 to 2017. For the analysis, the internationalization strategies available in the German dairy industry were first analyzed on the basis of an extended version of Perlmutter's (1969) EPRG model (Wind *et al.*, 1973). The panel data were estimated using Hausmann-Taylor estimation with clustered standard errors in order to cover endogeneity. Thereby we use the EBIT-margin as key figure for economic success, as it is widely used in literature (Li, 2007). Based on the results of the analysis and current studies, we derive recommendations for the internationalization of the German dairy industry.

Chapter VI (*"German Brazilian dairy supply chain integration: Prospects and ways"*, submitted to Agricultural Economics) analyzes whether an entry of German dairies into the Brazilian dairy market could provide a perspective for the future and what it could look like. Although Asia as a whole, and China in particular, are repeatedly mentioned with regard to opportunities for entrepreneurial growth, the Latin American milk market, one of the fastest growing markets worldwide, also offers companies interesting growth opportunities and the opportunity to launch new products on the market (Riley, 2017). First, we describe the characteristics of both milk production and the dairy industry in Brazil and Germany. The qualitative analysis is based on expert interviews conducted in both countries in the most important production regions for milk between November 2016 and January 2017. Furthermore, in a subsequent quantitative analysis we use the results from the paper presented in Chapter V in order to present a holistic picture of the potential of the Brazilian market for the German dairy industry. Building on this, we analyze where the opportunities and risks lie with regard to market entry. In addition, we identify the greatest challenges and give recommendations for action in dealing with them.

The dissertation concludes in Chapter VII with a short summary of important findings, some managerial and political conclusions and an outlook on possible future research directions.

References

Betzholz, T. (2010). Milcherzeugung vor dem Hintergrund der Kontingentierung. Monthly Statistical Bulletin Baden-Wuerttemberg, 9/2010.

BMEL (2019. Herstellungsmenge und Pro-Kopf-Verbrauch von Konsummilch 2018 weiter gesunken. Federal Ministery for Food and Agriculture, Berlin, available at: https://www.bmel-statistik.de/ernaehrung-fischerei/versorgungsbilanzen/milch-und-milcherzeugnisse/ (accessed 4 October 2019).

Bratschi, T., Feldmann, L. (2005). *Stomach Competence – Wachsen in gesättigten Märkten: Trends, Strategien und Konzepte für Lebensmitteleinzelhandel*, Food-Hersteller und Systemgastronomen. Deutscher Fachverlag, Frankfurt am Main..

Brokelmann, V., Gäde, S. (2017), „Branchenstudie Molkereiwirtschaft 2017“, HSH Nordbank.

Brümmer, B., Loy, J.-P. (2004). Einführung der Milchquotenbörse in Österreich. Expert opinion on behalf of the Federal Ministry of Agriculture, Forestry, Environment and Water Management, Vienna.

Brümmer, B., Spiller, A., Mehlhose, C. (2018). Der Markt für Milch- und Milcherzeugnisse im Jahr 2017. *German Journal of Agricultural Economics* 67: 119–137.

Bundeskartellamt (2015). Beschluss – In dem Verwaltungsverfahren B2-96/14. Federal Cartel Office of Germany, Bonn.

Chamber of Agriculture (2019). Nährstoffbericht für Niedersachsen 2017/2018. Chamber of Agriculture of Lower Saxony, Oldenburg.

Cornall, J. (2019). Arla commits to sustainable dairy sector in Nigeria. Dairy reporter.com, available at: https://www.dairyreporter.com/Article/2019/09/16/Arla-commits-to-sustainable-dairy-sector-in-Nigeria?utm_source=newsletter_daily&utm_medium=email&utm_campaign=16-Sep-2019&c=oVcyMhVABWGgxJntPJsv2qOnjlTKzGn0&p2 (accessed 29 September 2019).

Destatis (2019a). Employees and turnover in the German manufacturing industry from 2008 to 2018. Genesis-Online database, Federal Office for Statistics, available at: https://www-genesis.destatis.de/genesis/online (accessed 21 September 2019).

Destatis (2019b). Employees, turnover and invests in the German manufacturing industry from 2008 to 2017. Genesis-Online database, Federal Office for Statistics, available at: https://www-genesis.destatis.de/genesis/online (accessed 21 September 2018).

Destatis (2019c). Exports of German dairy and dairy products from 2008 to 2018. Genesis-Online database, Federal Office for Statistics, available at: https://www-genesis.destatis.de/genesis/online (accessed 21 September 2018).

Destatis (2011–2019d). Viehbestand und tierische Erzeugung 2016. Fachserie 3, Reihe 4, several vintages, Federal Office of Statistics, Wiesbaden.

dlz Agrarmagazin. Ausverkauf im Allgäu. Dlz Agrarmagazin, Deutscher Landwirtschaftsverlag GmbH, 7/2017: 124–127.

Eurostat (2019). Cows's milk collection and products – monthly data. Eurostat database, available at: https://ec.europa.eu/eurostat/de/data/database (accessed 23 September 2019).

Feurer, R., Chaharbaghi, K. (1994). Defining Competitiveness: A Holistic Approach. *Management Decision* 32: 49–58.

Friedrich, C. (2010). Milchverarbeitung und -vermarktung in Deutschland – eine deskriptive Analyze der Wertschöpfungskette. Thünen Working Papers, Institute of Farm Economics, Thünen-Institute, Braunschweig.

Grant, R.M., Nippa, M. (2006). *Strategisches Management.* Pearson Studium, Hallbergmoos.

Gries, T. and Hentschel, C. (1994). Internationale Wettbewerbsfähigkeit – was ist das? *Wirtschaftsdienst* 74: 416–422.

Guillouzo, R., Ruffio, P. (2005). Internationalization of European dairy co-operatives. *International Journal of Co-operative Management* 2(2): 25–35.

Handelsblatt (2011). Deutsches Milchkontor startet zum 1. Mai. Available at: https://www.handelsblatt.com/unternehmen/industrie/fusion-nordmilch-humana-deutsches-milchkontor-startet-am-1-mai/4116840.html?ticket=ST-16822871-0Chge6Eg3nhoAjUak4AF-ap1 (accessed 28 September 2019).

Heyder, M., Makus, C., Theuvsen, L. (2011). Internationalization and Firm Performance in Agribusiness: Empirical Evidence from European Cooperatives. *International Journal on Food System Dynamics* 2(1): 77–93.

Hörl, M., Hess, S. (2017). The Export Competitiveness of the European Dairy Industry. Paper presented at the XV Congress of the European Association of Agricultural Economists, 28 August – 1 September, Parma, Italy.

Janze, C., Meyer, J., Winkel, C., Theuvsen, L., Schmidt, C. (2018). Konjunkturbarometer Agribusiness in Deutschland 2018. Ernst & Young, Stuttgart.

Janze, C., Meyer, J., Winkel, C., Huchtemann, J.-P., Weinrich, R., Schmidt, C. (2019). Konjunkturbarometer Agribusiness in Deutschland 2019. Ernst & Young, Stuttgart.

Jürgens, K., Fink-Keßler, A., Poppinga, O., Wohlgemuth, M. (2015). Wertschöpfung von Molkereien. MEG Milch Board, Göttingen.

Kirner, L. (2009). Auswirkung der vollständigen Implementierung des Health-Check auf landwirtschaftliche Betriebe. *Ländlicher Raum*, online journal of the Austrian ministry for Agriculture, Forestry, Environment and Water Conservancy, available at: https://www.bmnt.gv.at/land/laendl_entwicklung/zukunftsraum_land_masterplan/Online-Fachzeitschrift-Laendlicher-Raum/archiv/2009/kirner.html (accessed 20 August 2019).

Lassen, B. (2011). Milchproduktion in Deutschland und Europa nach der Abschaffung Liberalisierung – Abschätzung künftiger Entwicklungen mit unterschiedlichen analytischen Ansätzen. Dissertation, Georg-August-University Göttingen, Göttingen.

Li, L. (2007). Multinationality and performance: A synthetic review and research agenda. *International Journal of Management Reviews* 9(2): 117–139.

Lindena, T., Tergast, H., Ellßel Raphaela, Hansen, H. (2018). Steckbriefe zur Tierhaltung. Institute of Farm Economics, Thünen-Institute, Braunschweig.

Lutter, M. (2009). Strukturwandel in der Milchviehhaltung: Ergebnisse einer regional differenzierten Befragung. Thünen Working Papers, Institute of Farm Economics, Thünen-Institute, Braunschweig.

Martin, L., Westgreen, R., van Duren, E. (1991). Agribusiness Competitiveness across National Boundaries. *American Journal of Agricultural Economics* 73(5): 1456–1464.

Mehlhose, C., Hunecke, C., Spiller, A., Brümmer, B. (2019). Der Markt für Milch- und Milcherzeugnisse im Jahr 2018. *German Journal of Agricultural Economics* 68: 62–84.

Meyer, J., Theuvsen, L. (2017). Internationalisierungsstrategien deutsche Molkereiunternehmen. Milchtrends, available at: https://www.milchtrends.de/fileadmin/milchtrends/5_Aktuelles/17-09_Internationalisierungsstrategien_Deutscher_Molkereiunternehmen.pdf (accessed 20 September 2019).

MIV (2019a). Wohin die Milch fließt. Dairy Industry Association of Germany, available at:

https://milchindustrie.de/wp-content/uploads/2018/11/Wohin-die-Milch-flie%C3%9Ft-2018.pdf (accessed 22 September 2019).

MIV (2019b). Exporte der deutschen Milchwirtschaft nach Zielregionen in 2018. Dairy Industry Association of Germany, available at: https://milchindustrie.de/wp-content/uploads/2017/10/Weltkarte-Deutsche-Exporte-Zielregion_2018.pdf (accessed 23 September 2019).

OECD, FAO (2019). OECD-FAO Agricultural Outlook 2019–2028. OECD Publishing, Paris.

Perlmutter, H.V. (1969). The Tortuous Evolution of the Multinational Corporation. *Columbia Journal of World Business* 4: 9–18.

Porter, M. (1980). *Competitive Strategy: Techniques for Analyzing Industries and Competitors*. Free Press, New York.

Riley, S. (2017). Milking the Latin American dairy market. Available at: https://www.dairyreporter.com/Article/2017/09/21/Milking-the-Latin-American-dairy-market (accessed 23 September 2019).

Schlippenbach, V., Pavel, F. (2011). Konzentration im Lebensmitteleinzelhandel: Hersteller sitzen am kürzeren Hebel. *DIW Wochenbericht* 78(13): 2–9.

Tangermann, S., Cramon-Taubadel, S. (2013). Agricultural policy in the European Union: An overview. Working paper No. 1302, Department of Rural and Agricultural Development, Georg-August-University Göttingen, Göttingen.

Theuvsen, L., Janze, C., Heyder, M. (2010). Agribusiness in Germany 2010. On the Way to New Markets, Ernst & Young, Stuttgart.

Top agrar (2019). Milchproduktion – Einige sind in die falsche Richtung unterwegs. Interview mit Peter Stahl, top agarar 8/2019.

Trabold, H. (1995). Die internationale Wettbewerbsfähigkeit einer Volkswirtschaft. *Vierteljahreshefte zur Wirtschaftsforschung* 64(2):169–185.

Trademap (2019). Data of trade for product groups 0401-0406. Inetrnational Trade Center, available at: https://www.trademap.org/Index.aspx (accessed 20 June 2019).

Utzig, S. (1987). Internationale Wettbewerbsfähigkeit, Unternehmensorganisation und Ordnungspolitik. *Wirtschaftsdienst* 67(8): 417–422.

Vitaliano, P. (2016). Global Dairy Trade: Where Are We, How Did We Get Here and Where Are We Going? *International Food and Agribusiness Management Review* 19(B): 27–36.

Vollrath, T.L. (1991). A Theoretical Evaluation of Alternative Trade Intensity Measures of Revealed Comparative Advantage. *Weltwirtschaftliches Archiv* 127(2): 265–280.

Weindlmaier, H. (2000). Die Wettbewerbsfähigkeit der deutschen Ernährungsindustrie: Methodische Ansatzpunkte zur Messung und empirische Ergebnisse. *Schriften der Gesellschaft für Wirtschafts- und Sozialwissenschaften des des Landbaues e.V.* 36: 239–248.

Wind, Y., Douglas, S.P., Perlmutter, H.V. (1973). Guideline for Developing International Strategies. *Journal of Marketing*, American Marketing Association 37(2): 14–23.

ZMB 2012. ZMB Jahrbuch Milch 2012. Zentrale Milchmarkt Berichterstattungs GmbH, Berlin

II Intensive Dairy Farming in Northern Germany: Development and Impact of the New Fertilizer Act

Authors: Johannes Meyer, Ludwig Theuvsen

Published in: AGRARIAN PERSPECTIVESS XXVI – Proceedings of the 26th Conference, pp. 219–225

Abstract

Germany, the largest milk producer in the European Union, increased milk production by 15.1% from 2007 to 2015. In 2015 intensive production regions with a milk production of more than 2,000 kg/ha contributed 59.8% to the German milk production whereby they represented 69.9% of the growth in German milk production from 2010 to 2015. The area of grassland and area of maize silage have a high positive and significant correlation with milk production. The regression model (R^2: 0.832) indicates the strong influence of these variables on milk production. Indeed data of twelve farms from an intensive production region show that fast grown farms get to limits due to the greening regulations on crop rotation and the nitrates limit of 170 kg/ha from manure of animal origin of the New Fertilizer Act. This will lead to increasing demand for land and the need of exporting nutrients to less intensive regions which will increase the costs of milk production.

1 Introduction

The European Union is the largest milk producer in the world. In 2015, European milk production amounted to 162.6 m tons, which is 24.4% of world milk production. Germany is the largest milk producer in the European Union with a production of 32.7 m tons in 2015, or 20.1% of European milk production (ZMB, 2016). German milk production decreased very slightly by 12.232 tons in 2016. That corresponds to a decrease of 0.04% (BLE, 2017). This was the first time in more than 10 years that German milk production did not increase over the preceding year (ZMB, 2011; ZMB 2016; BLE 2017). Overall, Germany's milk production grew by 15.1% between 2007 and 2015, which is above the EU-27 average of 9.5% (ZMB, 2011; ZMB, 2016). It should be noted, however, that milk production is not homogeneously distributed throughout Germany, as can be seen in figure 1.

Figure 1. Intensity of milk production per ha and year in Germany at county level in 2015

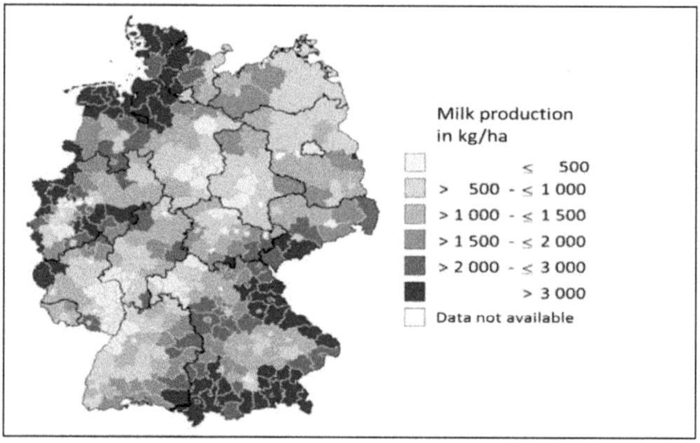

Source: Thünen-Institute, 2016.

Milk production is especially located where natural site conditions allow no or just a few types of land use beside grassland. This results in low opportunity costs for land, compared to regions where more numerous and more profitable opportunities for arable farming exist (Gömann *et al.*, 2006). Thus, the main milk production areas are located in regions with high shares of grassland. These can be found along the North Sea coast, in the upland regions in the middle of Germany, in the alpine uplands and along the Czech border. Further regions where intensive dairy farming takes place but where good conditions for arable farming also exist are the region bordering the Netherlands and the

Lower Rhine region (Lassen *et al.*, 2009). During the last decade, Germany's intensive dairy farming regions have attracted additional production volumes, whereas milk production has been shrinking in many less intensive areas (BLE, 2011–2016).

Intensive dairy farming regions are characterized by large quantities of manure. Similar effects occur in regions with intensive pig and poultry production, were large amounts of manure are also produced (Chamber of Agriculture, 2017). Due to the resulting nutrient surpluses, intensive livestock farming is held responsible for a substantial impact on nitrate contamination of groundwater (Taube *et al.*, 2013). Next to nitrate, phosphate is also an important water and environmental pollutant because it has the highest eutrophication potential (BMELV, 2013).

The European Union's Nitrates Directive is one of its main efforts to reduce nitrate leaching from agriculture and sets forth a number of measures (Velthof *et al.*, 2013). The Nitrates Directive helps to fulfil the goals of the European Union's Water Framework Directive, which was implemented in 2000 (BMU, 2013). The Directive of Utilization of Fertilizers, Soil Excipients, Culture Substrates and Plant Aids – known as the Düngeverordnung (DüV), or Fertilizer Act – regulates the utilization of fertilizers, including technical aspects of utilization and amounts. The act is the main instrument for implementing the European Union's Nitrates Directive in Germany (BMELV, 2013). The new regulations of the Fertilizer Act were passed by the German Federal Parliament on 31 March 2017 (BMEL, 2017a). The new regulation will come into force in the next planting season, starting after the 2017 harvest, and will affect dairy farms, among others. It regulates the balance of nitrates and phosphates as follows: nitrate use is limited to 60 kg/ha for a three year period, decreasing to 50 kg/ha in 2020, and phosphate use is limited to 20 kg/ha for a three year period, decreasing to 10 kg/ha in 2023. Furthermore, it introduces a lengthened blocking period for manure application, thus increasing the need for storage capacity to nine months for farms with more than three grazing livestock units per ha until 2020 and limiting the use of nitrogen from manure of animal origin to 170 kg/ha (BMEL, 2017b). In Lower Saxony, the N accumulation per ha will increase from 99.3 kg/ha to 123kg/ha of farmland due to the act's new regulations. This will lead to N accumulation close to the 170 kg N/ha limit or above in intensive livestock regions (Chamber of Agriculture, 2017). In the Netherlands, one of the most intensively farmed countries in the world, the European requirements are estimated to cut the dairy herd by about 160,000 animals, or 6.6% of the Dutch dairy herd, in the short run. In the long run, it is estimated that a further expansion of dairy herd and milk production will occur due to improved efficiency within the sector (USDA, 2017).

Against the background described above, the main objectives of this study are (1) to analyze the development of German milk production and its concentration in the time period 2010 to 2015, (2) to illustrate the coherence and strength of influence of the area of grassland and the area of silage maize production on milk production in Germany and (3) to examine the impact of the New Fertilizer Act on the production costs of dairy farms in an intensive region.

2 Materials and Methods

The study is based on official data from the Federal Office for Agriculture and Food on milk production at the county level from 2010 to 2015. Data for 2016 at the county level are not yet available. Additional data on the area of grassland and of silage maize production, disaggregated to county level, based on the agricultural structure survey of 2010 are provided by the KWS SAAT SE, a seed company based in Einbeck, Germany.

To analyze the concentration and development of milk production in Germany, descriptive statistics were used on county-level data. Concentration was measured as the share of milk production in intensive production regions relative to overall milk production in the respective state and in the country as a whole. Intensive production regions were divided into regions with moderate intensity, or milk production of 2,000 to 3,000 kg/ha, and with high intensity, or production of more than 3,000 kg/ha.

To do correlation and regression analysis, the data were analyzed on a normal distribution using the Kolmogorov-Smirnov test. The coherence of variables without normal distributions—milk production (tons), area of grassland (ha) and area of silage maize production (ha)—was evaluated with a correlation analysis, using the Spearman correlation coefficient.

The influence of grassland and silage maize on milk production was analyzed with an ordinary least squares model (OLS) using the software SPSS 23. Here, the area of grassland (A) and the area of silage maize (B) are the independent variables, influencing the dependent variable milk production in tons (Y). The correlation coefficients $\beta 1$ and $\beta 2$ estimate the influence of the independent variables. The residue item (U) shows the influence of other factors on the dependent variable.

$$Y = \beta_0 + \beta_1 * A + \beta_2 * B + Ui \tag{1}$$

To analyze the impact of the New Fertilizer Act on the production costs of dairy farms, data from 12 dairy farms in Northern Germany were collected with the help of a specialized consulting company in autumn 2016. The data refer to the financial year 2014/2015. This means that they show the economic situation from 1 July 2014 to 30 June 2015. Production costs were calculated with reference to the full cost accounting method. Therefore, opportunity costs were scheduled for production factors such as unpaid family workers and land. Depreciations were adjusted to the actual operating life expectancy.

3 Results and Discussion

In 2010 moderately intensive dairy farming regions accounted for 15.5% of German milk production, whereas highly intensive production regions were responsible for 43.3%. Thus, intensive production regions contributed 58.8% to German milk production in 2010. By 2015 the share of German milk production in intensive production regions had increased slightly to 59.8% including a slightly increased share in the highly intensive regions of 44.3% and a constant share in the moderately intensive regions of 15.5%. Although the share of the intensive regions in total milk production increased only by 1 percentage point, they play a very important role in the development of German milk production overall. In 2010 the intensive production regions were responsible for 76.2% of the surplus in milk production; the highly intensive regions alone contributed 62.4%. In 2015 the highly intensive regions were responsible for 67.8% of the increase in production, whereas the moderately intensive regions contributed only 10.7%. From 2010 to 2015 intensive production regions represented 69.9% of the growth in Germany's milk production, of which 54.9% was observed in highly intensive and 14.9% in moderately intensive regions.

The German states of Bavaria (24.9%), Lower Saxony (21%), North Rhine-Westphalia (10.2%) and Schleswig-Holstein (9.1%) represented 65.1% of German milk production in 2015. Milk production of more than 2,000kg/ha also occurs in Hesse, Rhineland Palatinate, Baden Wuerttemberg, Saxony, Brandenburg and Thuringia.

Table 1. Contribution of intensive dairy regions to total milk production in different states (%)

State	2010	2011	2012	2013	2014	2015	Average
Schleswig-Holstein	92.4	92.7	92.2	92.3	92.5	92.3	92.4
Lower Saxony	77	77.9	77.8	77.8	77.9	78.3	77.8
North Rhine-Westphalia	67.1	67.5	67.5	67.8	68.2	68	67.7
Hesse	18.2	18.4	18.9	18.8	n.a.	18.6	18.6
Rhineland Palatinate	51.5	51.5	51.2	51.7	51.7	51.4	51.5
Baden Wuerttemberg	41.4	41.3	41	41	41.1	41.3	41.2
Bavaria	75.3	75.5	75.9	75.7	76.3	76.6	75.9
Brandenburg	1.4	1.5	1.4	0	1.5	1.5	1.2
Saxony	58	58	58.1	57.9	58.2	59	58.2
Thuringia	23.7	23.9	24.2	24.5	24.9	25.3	24.4

n.a. = not available

Source: Authors' own calculation based on BLE, 2011–2016.

As can be seen in Table 1, there are great differences regarding the role of intensive production regions in the various German states. Milk production is most concentrated in Schleswig-Holstein, where 92.3% of milk production came from intensive production regions in 2015. Schleswig-Holstein is followed by Lower Saxony (78.3%), Bavaria (76.6%) and North Rhine-Westphalia (68%). The lowest concentration of milk production can be seen in Thuringia (24.4%), Hesse (18.6%) and Brandenburg (1.5%). In all German states, the highly intensive regions represent higher shares of total milk production than the moderately intensive regions. In Schleswig-Holstein the highly intensive regions contributed 70.3% to the state's total milk production in 2015. The highly intensive regions' share of total milk production was 67.6% in Lower Saxony, 55.5% in Bavaria and 51.8% in North Rhine-Westphalia. The growing concentration of milk production can also be seen when examining production growth in intensive regions. In Lower Saxony, for example, the intensive production regions were responsible for 91.1% of production growth in 2015, with 78% deriving from highly intensive regions. In some states, intensive regions contribute more than 100% to production growth. This is because production grew in these regions, while the states' overall milk production decreased. This is the case in Rhineland Palatinate (198.5%) and Saxony (136.4%).

Available grassland is an important factor for milk production. The weighted average share of grassland on total farming land in highly intensive regions is 49.2%. Moderately intensive regions have a weighted average share of 27.3%, and non-intensive regions 22%. The area of silage maize production is also higher in intensive dairy farming regions. In highly intensive regions, the weighted average share of arable land dedicated

to silage maize is 37.6%, and the weighted average share in moderately intensive regions is 22.2%. In other regions silage maize accounts for a share of 15.2% of arable land. The share of arable land dedicated to silage maize production is an important determinant of further growth in milk production because due to the greening regulations, the share arable land dedicated to the main crop is limited to 75% (Chamber of Agriculture 2016). In a few German counties, the share dedicated to silage maize is already up to 75%. Though at county level that share may average 37.6% in highly intensive regions, at the individual farm level it can be much higher, as is the case on the farms we analyzed in the Northern German intensive region, where it is already nearly 75%. The close relationship between milk production (tons) and the areas of grassland (ha) and silage maize production (ha) can be seen when analysing the correlation. Both factors have a very strongly positive and highly significant correlation with milk production, as can be seen in Table 2.

Table 2. Correlation matrix with Spearman correlation coefficients

	Milk production	**Area of grassland**	**Area of silage maize**
Milk production	1	0.889**	0.887**
Area of grassland	0.889**	1,000	0.766**
Area of silage maize	0.887**	0.766**	1

**significance level 0,01

Source: Authors' own calculation

The regression model analyzed the influence of grassland area (A) and the area of silage maize (B) on German milk production. As can be seen in equation 2, both factors have a highly positive and significant influence.

$$Y = -9{,}448{,}894^{1} + 4{,}935^{1} * A + 5{,}647^{1} * B + Ui \quad (2)$$

[1]significance level 0.01

The coefficient of determination, R^2, is 0.832; this means that 83.2% of the estimated residues can be explained by the independent variables "area of grassland" and "area of silage maize", both of which are essential for milk production. The result confirms the theory that milk production is allocated where natural resources allow no or just a few types of land use other than grassland. Furthermore, it indicates the strong influence of silage maize production on milk production in the prevailing production systems in Germany.

Another point that will influence the development of milk production is the new Fertilizer Act. The data from German dairy farms in intensive regions show that the farms which have grown strongly over the last years will not fulfil the tighter regulations of the new Fertilizer Act because their nitrate levels from animal manure are above the now more strictly defined 170 kg/ha limit. The share of total farm land dedicated to grassland on these farms is 45.9% Therefore, the farms will be impacted by the abolishment of the derogation option, which allowed them to put 230 kg/ha of nitrates from animal manure on grassland. The farms in our study have a surplus of 19.9 kg N/ha (Niemann, 2016). As a result, they will be forced to rent additional land, export the nitrate surpluses to regions with lower livestock densities, reduce their herd size or reorganize their farms by, for instance, outsourcing of the rearing of calves. Export opportunities are restricted due to competition from pig and poultry farmers in intensive livestock production regions, who are also affected by the new legislation. These farms are further affected by the stricter phosphate limit because of the higher amounts of phosphates produced, especially in pig manure (Chamber of Agriculture, 2017). This situation will lead to an increased demand for land even though land prices are already high in the intensive regions we studied. The opportunity costs for land are currently about €600/ha p.a. for arable land (Niemann, 2016). How land prices will develop is not completely clear. In general, prices increase when demand increases, as seen in the past. While rental prices for land are much higher even than €600/ha p.a. in other intensive livestock regions, they have stayed constant or even decreased in some intensive dairy regions because of farm failures due to the low milk prices in recent years. Nevertheless, recent increases in land prices have led to an increase in production costs of 0.7 ct./kg energy corrected milk (ECM). The costs for the export of manure within the blocking period will result in additional costs of 0.4 ct./kg ECM. Here cost of storage of €5/m^3 were calculated and an average transport distance of 30 km is assumed (Niemann, 2016). Indeed, these costs may increase in future due to the fact that the legally stipulated nitrate balance of 60 kg/ha will decrease to 50 kg/ha in 2020. Furthermore, export distances are also likely to increase due to competition with livestock farms and substrate from biogas plants because the new Fertilizer Act also limits the application of substrate to 170 kg N/ha (BMEL, 2017a). Last but not least, the low prices for mineral fertilizers could decrease the willingness of arable farmers to use manure.

4 Conclusions

From 2010 to 2015, German milk production grew more strongly than the European average, whereby the growth was mainly concentrated in intensive dairy farming regions. From 2010 to 2015, intensive production regions were responsible for 69.9% of production growth in Germany, whereby the highly intensive regions accounted for 54.9%. The result of this analysis confirms the theory that milk production is heavily based on grassland due to its low opportunity costs. Furthermore, the study confirms the strong influence of area of grassland and area of silage maize on milk production. Especially on farms which have strongly increased their herd sizes, EU greening regulations will further limit the growth of milk production. The new Fertilizer Act will increase the production costs of dairy farms, especially those of farms which grew quickly in the past. Affected dairy farms will have to develop strategies to solve this problem. Cost analyses show that strategies such as the outsourcing of the breeding of calves can decrease the surplus of nitrates to below the 170 kg/ha limit at the farm level (Niemann, 2016).

Nevertheless, in most cases this will end up in increasing production costs on affected dairy farms, which will reduce their competitiveness against foreign competitors, especially those farms in less intensive areas with low opportunity costs. Indeed, there are competitors such as farms in the Netherlands which are more heavily impacted by the implementation of the European regulations. Policymakers should deal with the problem of nutrient surpluses by supporting the development of solutions, for example, economically more attractive export systems for nutrients into less intensive regions or more nutrient-efficient ratios for animals (Kröger, 2016). These measures will help remedy the impact of the Fertilizer Act on dairy farming in Germany and preserve its competitiveness since, in the end, there are many good reasons besides site conditions, such as farmers' knowledge and presence of necessary infrastructure, why dairy farming is located in these areas and has been further concentrated there.

References

BLE (2011–2016). Milcherzeugung und –verwendung bis auf Kreisebene in Deutschland. Federal Office for Agriculture and FoodBLE 2017. Kuhmilchlieferung der Erzeuger an deutsche milchwirtschaftliche Unternehmen. Federal Office for Agriculture and Food.

BMEL (2017a). Verordnung zur Neuordnung der guten fachlichen Praxis beim Düngen. Federal Ministry of Food and Agriculture, available at: http://www.bmel.de/SharedDocs/Downloads/Service/Rechtsgrundlagen/Entwuerfe/EntwurfDuengeverordnung.pdf?__blob=publicationFile (accessed 10 April 2017).

BMEL (2017b). Strengere Regeln für die Düngung. Federal Ministry of Food and Agriculture, available at: https://www.bmel.de/DE/Landwirtschaft/Pflanzenbau/ /Ackerbau/Texte/DuengepaketNovelle.html (accessed 25 April 2017).

BMELV (2013). Novellierung der Düngeverordnung: Nährstoffüberschüsse wirksam begrenzen. Ministry of Diet, Agriculture and Consumer Protection, special issue 219.

BMU (2013). Die Wasserrahmenrichtlinie. Eine Zwischenbilanz zur Umsetzung der Maßnahmenprogramme 2012. Ministry of Environment, Nature Protection and Reactor Safety.

Chamber of Agriculture (2016). Merkblatt zum Greening 2016. Chamber of Agriculture of North Rhine-Westphalia, available at: https://www.landwirtschaftskammer.de/foerderung/formulare/merkblaetter/mb-sammelantrag-2016-greening.pdf (accessed 10 April 2017).

Chamber of Agriculture (2017). Nährstoffbericht in Bezug auf Wirtschaftsdünger für Niedersachsen 2015/2016. Chamber of Agriculture of Lower Saxony.

Gömann, H., Kreins, P., Zabel, A. (2006). Wohin wandert die Milchproduktion in Deutschland? *Agricultural Research Völkenrode*, special issues 299: 97–108.

Kröger, R. (2016). Gülleseparation und Güllefeststoffvergärung: Akzeptanz und Wirtschaftlichkeit. Dissertation at the Chair of Agribusiness Management, Georg-August-University Göttingen, Göttingen.

KWS (2017). Areas of Farming, Arable, Grassland, Corn and Maize on County Level in Germany 2010, Data of the agricultural structural survey 2010, provided by the KWS SE.

Lassen, B., Isermeyer, F., Friedrich, C. (2009). Regional changes in German dairy production. *Agrarwirtschaft* 58, booklet 5/6: 238–247.

Niemann, F. (2016). Novellierung der Düngeverordnung: Anpassungsstrategien von Milchviehbetrieben in einer Intensivregion. Master Thesis, Georg-August-University Göttingen, Göttingen.

Taube, F., Schütte, J., Kluß, C. (2013). Auswirkungen der Berücksichtigung von Gärresten auf den Anfall organischer Dünger in einer novellierten Düngeverordnung, dargestellt am Beispiel Schleswig-Holstein. *Berichte über Landwirtschaft*, special issue 219.

Velthof, C.L., Lesschen, J.P., Webb, J., Pietrzak, S., Miatkowski, Z., Pinto, M., Kros, J., Oenema, O. (2014). The impact of the Nitrates Directive on nitrogen emissions from agriculture in the EU-27 during 2000–2008. *Science of the Total Environment* 468–469: 1225–1233.

Thünen-Institute (2016). Focus Areas of Milk Production in Germany in 2015. Thünen-Institute, Braunschweig.

USDA (2017). New Phosphate Reduction Plan Sets Limits to Dutch Dairy Production. Gain Report, United States Department of Agriculture, available at: https://gain.fas.usda.gov/Recent%20GAIN%20Publications/New%20Phosphate%20Reduction%20Plan%20Sets%20Limits%20to%20Dutch%20Dairy%20ProductionThe%20HagueNetherlands2-9-2017.pdf (accessed 15 April 2017).

ZMB (2011–2016). ZMB Jahrbuch Milch 2011–2016. Zentrale Milchmarkt Berichterstattung GmbH.

III Drivers of Large Herd Sizes in German Dairy Farming: Development and Outlook

Authors: Johannes Meyer, Jan-Henning Feil, Christian Schaper

Submitted to: Outlook on Agriculture

Abstract

Like the entire agricultural sector in Germany, dairy farming is characterized by progressive structural change, leading to a declining number of farms and an increasing number of cows per farm. This development affects all herd size classes except that of more than 100 animals. In this herd size class, both the number of farms and the number of cows have increased continuously during the period under review from 1992 to 2018. Due to the capital intensity, structural changes in agriculture and thus also in dairy farming are generally accompanied by long-term investments. In the past, milk production in Germany has been significantly influenced by the Common Agricultural Policy of the European Union (CAP), which has, however, been increasingly liberalized (CAP). However, other factors such as capital and opportunity costs, prices and the mood in the sector have also changed considerably over time. This paper examines both qualitatively and quantitatively how individual factors affected the development of the number of cows in the large dairy herds in the period from 1992 to 2018 and, building on these results, derives possible developments for the future. The results show a clearly negative influence on the development of this herd size by the milk quota: Against that, it can be seen that the lower capital costs, caused by the historically low interest rates for agricultural long-term loans in particular had a positive influence on it. However, with a view to the future, other factors are likely to have a limiting effect after the end of the milk quota.

1 Introduction

Farmers have normally just a few options to improve their revenues in terms of increased prices. Therefore they normally try to reduce production costs in order to get ceteris paribus higher revenues. There is evidence that larger herd sizes lead in general to lower production costs for milk, resulting from economies of scale, but also of the ability to use technical progress, and so on (MacDonald *et al.*, 2007; IFCN, 2013). Due to the regulations of the European Common Agricultural Policy (CAP) including the milk quota which was introduced in 1984, growth of farms and herds had been strongly restricted for a long time. This has changed with an ongoing liberalization of the CAP (Offermann *et al.*, 2002; Tangermann and Cramon-Taubadel, 2013).

Since the beginning of 1992 to 2018, the number of dairy farms in Germany has fallen continuously from 236,000 to 61,087 units while the number of cows in Germany is also declining. From 5.4 million head in 1992, the number fell to just under 4.1 million in 2018. Thereby the average number of cows per farm rose from 22.7 in 1992 to 66.6 in 2018. (Lutter, 2009; ZMB, 2012; Destatis, 2019).

With regard to the structural change in dairy farming, it can be seen that the number of animals and holdings in the largest herd size class with 100 and more cows is increasing over the entire period. Even in the second-largest herd size class with 50 to 99 cows, the number of animals and holdings has fallen in recent years, especially after the introduction of the expansion of the milk quota as part of the abolition of this (ZMB, 2012; Hörl and Hess, 2017; Destatis, 2010,2018). Studies show that, among other things like, the initial farm size has an influence on the growth and survival of farms (Weiss, 1999). Based on this and the assumption of "survival analysis" (Stigler, 1958), which assumes that farms invest in structures with which they can realize the lowest costs, with the given technologies and input prices at the time of the investment, we focus our analysis on the development of large herds with more than 100 animals (McDonald *et al.*, 2017).

Due to the capital intensity of agriculture, structural developments and farm growth are mostly accompanied by long-term investments (Weiss, 1999; DBV, 2019; Koester and Cramon-Taubadel, 2019). This also applies to dairy farming as its mainly based on stable production systems in Germany (Lindena *et al.*, 2018). Beside stables, investments linked to milk production are made in form of milking systems, feeding systems, ancillary equipment, etc. (Sauer and Lactaz-Lohmann, 2015; Dorfner and Hofmann, 2019).

In the following, we will examine the theoretical background in Chapter 2 with regard to the causes for the development of large herds in German dairy farming and, building on this, derive our research hypotheses. In Chapter 3 we describe the development of the structure in German dairy farming in detail, as well as those of the identified influencing factors. In the fourth chapter we describe the data basis and methodology. Chapter 5 presents and discusses the results. Building on this, we finish with an outlook and conclusions in Chapter 6.

2 Theoretical Background

Like agriculture as a whole, milk production is strongly influenced by European agricultural policy. The best known agricultural policy instrument is the milk quota. With the introduction of basic regulation for the common milk market 804/68 of the European Economic Community (EEC), the German milk market and thus also German milk production has been influenced and controlled essentially by the European agricultural policy. The system has been repeatedly revised over time (Kreß, 2005).

On the one hand, agricultural policy was gradually liberalized, including in the dairy sector. While this was largely left out in the Mc-Sharry-Reform of 1992, milk prices approached the intervention price level as a result of concessions in the WTO negotiations in the mid-1990s, in the form of limited export refunds, limited internal support and the opening of markets to third countries (Kreß, 2005). In 2003 it was decided to extend the milk quota system until 2014/15 and to increase the quotas in three stages by 0.5% per year from 2006 onwards. Due to the strong increase of the milk price, the quotas were increased by 2% in 2008/09 (Kreß, 2005; Dialer *et al.*, 2010). At the so-called health check in 2008, it was decided to phase out the quota system by 2015. In order to make a smooth transition, it was decided to increase the milk quotas in Europe by 1% from 2009/10 to 2013/14 (Kirner, 2009). On 31st of December 2015 the milk quota was finally abolished (Hörl and Hess, 2017).

On the other hand, the liberalization applies to the tradability of quota. From the end of 1993, quotas could also be transferred without farms or areas, but certain preconditions had to be met, such as a transfer of at least two years and to an active milk producer. In 2000, tradability was further facilitated by the abolition of area binding and the introduction of the milk exchange, although trade remained regionally restricted (at the level of the federal states or administrative districts) (Höller, 2015; Brümmer and Loy, 2004).

From mid-2007, the regional restrictions were further relaxed. Milk quotas could be traded in the "Übertragungsgebiet West" in the old federal states and in the "Übertragungsgebiet Ost" in the new federal states (Betzholz, 2010).

Another factor influencing investments and thus the development of large dairy herds are factor price relations. These relations change both through a change in factor prices outside agriculture and through the introduction of new technologies both inside and outside agriculture. Since economic growth in an economy is associated with rising wages, the opportunity costs for workers rise, while the costs for capital ceteris paribus fall relatively. Accordingly, the optimal combination of labour shifts in favour of capital (Koester and Cramon-Taubadel, 2019). As others, also agricultural firms rely on debt capital to finance operations, investments, modernization and so on. Easily available credit has favored many of the major long-term changes such as farm growth, specialization and technical progress in the agricultural sector (Barry and Robison, 2001). In the case of debt capital, the division into long-term and short- or medium-term debt capital is particularly important in agriculture. Short- and medium-term loans with a term of less than one to five years are used for consumption purposes, the purchase of working capital and to bridge short-term liquidity bottlenecks. In contrast, farmers use long-term loans to finance long-term investments, usually including loans to finance stables (Koester and Cramon-Taubadel, 2019; DBV, 2019). Dairy farming in Germany is mainly based on stables. In 2010, 72% of cows were kept in pit stables, 27% in tethered stables (Lindena *et al.*, 2018; Bundesinformationszentrum Landwirtschaft, 2019).

In terms of investments of farmers, the output price is an important variable as the payout price is an essential factor for net return per kg of milk. Thus, ceteris paribus, this also has a positive effect on the financial situation of farmers which has a considerable influence on long-term investments in land and buildings (Vasavada and Chambers, 1986; Chavas and Magand, 1988; Elhorst, 1992; Rougoor *et al.*, 1996).

Some studies point out that, with regard to the explanation of the investment behavior of farmers, there are variables besides variables of neoclassical production theory that have a considerable explanatory content but are difficult to quantify (Elhorst, 1992; Jacobsen, 1996). Against this background, we have included the assessment of the business situation and business expectations in the milk processing industry of the ifo-Institute, as we do not have comparable data from milk production over the period under review and in order to cover the mentioned shortcomings.

Opportunity costs are another important factor influencing investments in dairy farming and thus the development of larger herd sizes. Studies show that milk production is expanded where opportunity costs are low. This is especially true in low prices or under difficult economic conditions, as factor mobility is higher in regions with alternative uses. Classic dairy cattle regions in Germany have a high proportion of grassland, as the opportunity costs are low due to the limited alternatives (Chavas Magand, 1988; MacDonald *et al.*, 2007; Lassen *et al.*, 2008).

Based on these findings we derive the following hypotheses:

- The quota has negative effect on the development of large herds and thus negative effects on investments.
- The milk price has positive effect on the development of large herds in German dairy production.
- The cost of capital has influence on the investments.Due to the historically low cost and therefore higher opportunity costs for labor, we hypothesize that the low interests have a positive effect on the development of large herds.
- The assessment of the business expectation and the assessment of the current business situation play a decisive role in the development of large herds in German milk production.
- Opportunity costs in form of higher grain prices have a negative impact on the development of large herds.

3 Development of the Structure of Dairy Farming in Germany and Considered Factors

3.1 Structure of Dairy Farming

The structure of German milk production has changed considerably during the period under review as can be seen in Figure 1. The number of cows in the small herd size classes with one to nine, 10 to 19 and 20 to 49 cows has declined continuously over the considered period from 1992 to 2018. As a result, their share on the total number of cows kept in Germany felt from 74.1% in 1992 to 18% in 2018.

Figure 1. Development of herd sizes and farms in German dairy farming

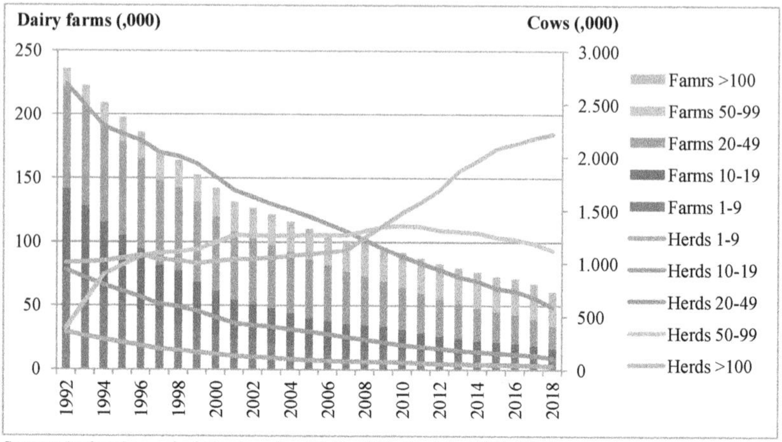

Source: Authors' own depiction based on ZMB, 2012; Destatis, 2011–2018.

The number of animals in herds with 50 to 99 animals initially rose sharply from the beginning to the middle of the 1990s. Between 1992 and 1996, the number of cows increased from 387,000 to 1.1 million, which represents an increase from 7.2 to 20.5% of cows kept in Germany. As a result, the growth of this herd size slowed until it peaked in 2010 with almost 1.4 million cows (32.3%). From 2010 onwards, the number of cows kept in this herd size fell to 1.1 million animals by 2018, respectively 27.6% of cows kept in Germany.

In contrast, the number of animals in herds with more than 100 cows[1] increased over the whole period. The only exception is the period from 1996 to 1999, when the number of cows in this herd size class felt slightly. From just over one million cows in 1992, the number rose to 2.2 million cows in 2018, which means that the proportion of cows kept in Germany in this herd size class rose from 18.7% in 1992 to 54.5% in 2018. The increase is mainly due to the period from 2008 to 2018. During this period the number of animals increased from 1.2 million to 2.2 animals.

[1] The Federal Office for Statistics further subdivides the herd size class with 100 and more cows from 2003 on by introducing the herd size classes 100 to 199 animals, 200 to 299 animals and more than 300 animals. From 2013 onwards the herd size classes with more than 100 cows were divided into the size classes 100 to 199 animals, 200 to 499 animals and more than 500 animals. In 2018, 49.1% of the 2.2 million cows in herds with more than 100 cows were kept in the herd size class with 100 to 199 animals, 30.6% in the herd size class with 200 to 499 animals and 20.3% in the herd size class with more than 500 animals. Due to the short-term period and the repeated changes in the subdivision of herd size classes, we cannot take these further into account in our considerations and calculations and group them under the herd size class with 100 and more animals.

A similar development as for the number of animals in the different herd size classes exists for the number of farms by herd size classes. Of the total of 236,000 dairy farms in Germany in 1992, only 2,900 (1.2%) had herds of more than 100 cows. 11,300 farms (4.8%) had herds of 50 to 99 cows. The other 221,800 farms (94%) belonged to the small herd size classes. Until 2018 the number of farms in the small herd sizes felt to 34,300, corresponding to 56.2% of all dairy farms. The number of farms with a herd size of 50 to 99 cows rose from 11,300 (4.8%) in 1992 to 15,900 (26%) in 2018. However, the number of farms in this herd size class fell from 2008 onwards. By contrast, the number of farms with herd sizes of more than 100 cows rose continuously over the entire period to 10,900, respectively 17.8% of dairy farms in 2018.

Table 1. Development of number of cows in dairy herds with more than 100 cows in selected states

State / Number of cows (,000)	2010	2012	2014	2016	2018
Germany	1,415.2	1,628.1	1,943.2	2,123.0	2,227.1
Baden-Wuerttemberg	41.3	55.8	78.5	n.a.	112.5
Bavaria	55.6	84.9	108.9	159.5	196.6
Brandenburg	151.4	150.6	157.7	151.2	145.0
Hesse	30.0	41.6	53.3	59.8	64.0
Mecklenburg-Western Pomerania	158.4	164.4	171.9	169.1	160.0
Lower Saxony	290.2	367.7	477.2	556.6	590.8
North Rhine-Westphalia	122.6	154.8	202.2	226.9	245.1
Rhineland Palatinate	27.1	29.6	47.6	52.1	55.2
Saxony	161.9	163.1	168.4	167.2	162.1
Saxony-Anhalt	113.2	114.6	118.1	116.5	112.7
Schleswig-Holstein	146.4	188.5	232.0	256.3	273.9
Thuringia	101.1	99.3	104.0	101.8	95.9

Source: Authors' own depiction based on Destatis, 2010–2018

With regard to the development of the number of animals in the herd size class with more than 100 animals, there are clear regional differences within Germany (see Table 1). While in the old federal states the number of cows in herds with more than 100 cows rises sharply over time, in the new federal states the number rises only slightly until 2018 or even declines like in Thuringia and Saxony-Anhalt.

3.2 Interest Rate for Long-Term Loans and Agricultural Loans

Interest rates for long-term corporate loans fell significantly in the period under review. At first, interest rates for long-term corporate loans fell from over 11% in 1992 to 5.3% in 1999 (see Figure 2). By 2000, they had risen again to 7% before they felt to 4.3% in the end of 2005. In the following they rose to 5.6% by 2008. In the wake of the financial crisis, interest rates then fell to 1.8% by 2015. Since then, interest rates for long-term corporate loans have moved between 1.7 and a good 2% at a historically low level.

Against that, loans have increased almost continuously in the period under review from 1992 to 2018 in German agriculture and forestry. From 1992 to 2018, loans rose from € 25.8 billion to € 53.2 billion in 2018, which represents a total increase of € 27.4 billion or 106.4% over the entire period under review. However, the increase up to 2009 was significantly lower than that of the last ten years of observation. From 2009 to 2018, loans increased by € 19.8 billion, which corresponds to 69.8% of total growth.

Figure 2. Development of agricultural loans and interest-rate for long-term lonas

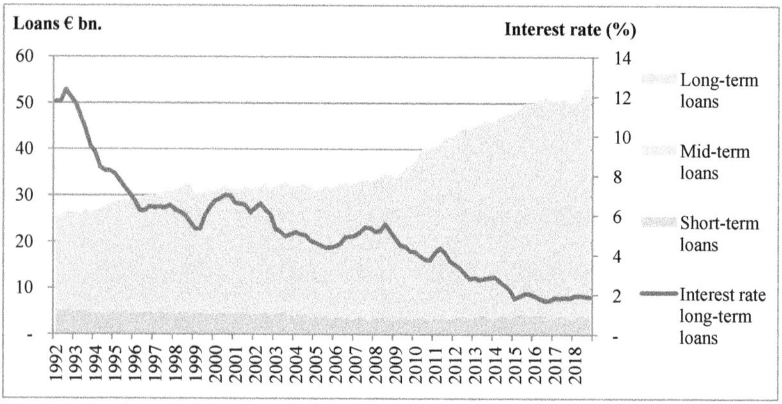

Source: Authors' own depiction and calculation based on Bundesbank, 2019a, b, c, f, g, h.

Long-term loans account for the largest share of total loans. In 1992 they accounted for € 18.4 billion or 71.4% of total loans in agriculture and forestry. By 2018, long-term loans had risen to € 45 billion, accounting for 84.7% of all loans. By the end of 2008, long-term loans had risen relatively moderately by an average of 2.6% per year. Starting in 2009, the total of long-term loans in agriculture and forestry grew much faster, averaging 5% per year. Against this, short and medium-term loans together hardly changed over time in absolute values. As a result, their share on total loans decreased from 28.6% in 1992 to 15.3% by the end of 2008.

Of the total growth of € 27.4 billion over the entire period, € 26.6 billion, or 97.2%, is attributable to long-term loans. By contrast, short and medium-term loans account for only € 0.8 billion, or 2.8%, of the growth in agricultural and forestry loans.

3.3 Assessment of Business Situation, Business Expectation and Milk Price

From 1992 to 2009, the assessment of business expectations was mostly less favorable with an average index of -8.3 points (see Figure 3). Also the assessment of the current business situation in the period under review was mostly poor, which is indicated by the average index value of -3.6 points from 1992 to 2009. Thereby the index for the business situation ranged from -31 to 23.3 points, and that of the business expectations from -39.5 to 13.6 points.

Figure 3. Development of assessment of business situation, business expectation and milk price

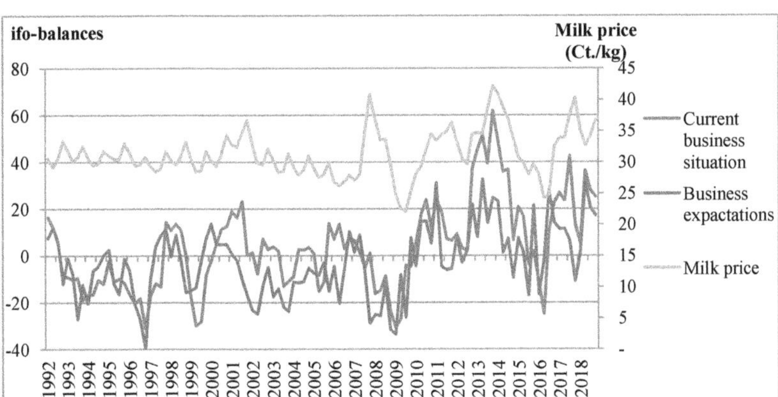

Source: Authors' own depiction based on ifo-Institute, 2019; European Commission, 2019.

From 2009 to 2018, the assessment of the business situation and business expectations improved significantly. During this period, the majority assessed the business situation as good and the business expectations as favorable. On average, the assessment of the business situation in the period from 2009 to 2018 was 16.5 points, that of the business expectations 6 points. The values for the assessment of the business situation in this period ranged from -30.6 to 61.9 points, those of the business expectations from -33.5 to 32.7 points.

The development of the milk price also differs greatly over time. In the period from 1992 to 2018 the milk price showed a relatively constant development with relatively even fluctuations. An exception to this is 2001, when the milk price rose to an average of 36.7 Ct./kg in the last quarter. Over the period from 1992 to 2007, the average milk price was 30.6 Ct./kg (see Table 2).

Table 2. Characteristics of the German quarterly milk price

(Ct./kg)	Minimum	Median	Average	Maximum	Standard deviation
Milk price 1992–2018	22.1	30.7	31.4	42.1	3.7
Milk price 1992–2007	26.3	30.4	30.6	36.7	2.0
Milk price 2007–2018	22.1	33.7	32.6	42.1	4.8

Source: Authors' own calculation based on European Commission, 2019.

From 2007 onwards, the characteristics of the milk price changed completely. The previous relative constant seasonality disappeared completely and volatility increased significantly. The standard deviation of the milk price in the period from 1992 to 2007 was 2 Ct./kg while it rose to 4.8 Ct./kg in the time from 2007 to 2018. However, the average milk price in the period from 2007 to 2018, at 32.6 Ct./kg was almost two cents higher than that of the period from 1992 to 2007 (30.6 Ct./kg).

4 Data and Methodology

4.1 Data

The data of the study is based on secondary data. The data of the number of cows in the herd size classes come from the Federal Office for Statistics. The data were collected every two years until 2003, annually from 2003 to 2011 and from 2011 onwards six-monthly. They were interpolated linearly on a quarterly basis. The data of the grain price index do also come from the Federal Office for Statistics.

The interest rates for long-term corporate loans[2] are taken from the statistics of the German Bundesbank. As there are only figures for the sum of loans used in agriculture and forestry and there is no more precise information on the amount of funds invested in milk production, we use these data only for qualitative and not for the quantitative anal-

[2] Due to changes in the survey, there are no uniform interest rates for loans to agricultural enterprises. The interest rate from 1996 to 2003 corresponds to the effective interest rate for long-term fixed-rate loans to enterprises and self-employed persons of € 500,000 to € 5 million. The interest rate from 2003 to 2018 corresponds to the effective interest rate on non-financial corporations up to € 1 million with an initial fixed interest period of more than 5 years. The interest rate from 1992 to 1996 was estimated on the basis of the difference between the discount and base rates and the effective interest rates due to the lack of data. The correlation between effective interest rates and discount and base rates is 0.928 ($p<0.01$). The approximate premium from 1992 to October 1996 is 3.75 percentage points.

ysis. The milk price comes from the EU Commission. It corresponds to the average milk price paid in Germany.

The data of business situation and business expectation comes from the ifo-Institute. The ifo data is collected monthly. The companies rate the current business situation as “good”, “satisfactory” or “bad”. Business expectations are rated by companies as “more favorable”, “unchanged” or “less favorable”. The ifo balances of the current business situation are calculated from the difference between the percentages of the answers “good” and “bad”. The balance of the business expectations therefore results from the difference between the percentages of the answers “more favourable” and “less favorable” (ifo Institute 2018).

4.2 Methodology

We assume that the independent variables milk price, interest rate for long-term loans, current business situation, business expectations and grain price at time t_0 have no influence on the dependent variable in form of the amount of cows (,000) in the herd size with 100 and more cows in t_0. Rather, production at time t_0 results from decisions made by the farmer in the past on the basis of the given information for price, interest rate, business situation, business expectations and opportunity costs in form of the grain price. The delay is due to the fact that a positive change in the number of animals in the considered herd size of more than 100 cows is accompanied by preliminary planning as cows are mostly kept in stables in Germany. Thus approval documents must be submitted, usually a bank must be looked for the financing takes over, which accompanies with a rating of the farmer, an architect must be found and so on and so forth. And at the end the project has to be approved and built. In order to take planning, approval and commissioning into account we have lagged the above variables by 2.5 years, respectively 10 quarters. Since the data on herd size classes are available from 1992 onwards and the data on independent variables from 1991 onwards, the period analyzed in the calculations shrinks due to the time lag of 2.5 years. Thus, in the analysis the period from the third quarter of 1993 onwards can be taken into account due to the delayed impact of the variables. In total we have 102 observations. To capture the effects of the milk quota we implemented a dummy for the quota. Furthermore we implemented a dummy to control for a time trend.

The time series was first analyzed in terms of stationarity, heteroskedasticity and autocorrelation. We have tested the variables with the Augmented Dickey Fuller test for stationarity (results see Table 1 in the appendix). Based on the results, we formed first differences for the variables interest rates and grain price to make them stationary as shown in equation 1 (Rottmann and Auer, 2010).

$$\Delta x_t = x_t - x_{t-1} \tag{1}$$

We used the Breusch-Godfrey Test to control for autocorrelation. To control for heteroscedasticity we used the Breusch-Pagan Test. We used an OLS regression model to estimate the impacts of the considered variables on the herd size. As we found heteroscedasticity and autocorrelation we estimate our regression using Newey-West Standard Errors which are heteroscedasticity and autocorrelation consistent (Croux *et al.*, 2003; Rottmann and Auer, 2010). To control for multicollinearity we used the variance inflation factors (VIF). As can be seen in Table 2 in the appendix the results imply that the model does not suffer from multicollinearity.

According to the mentioned assumption our model has the following specification as shown in equation 2.

$$Y_t = C + \beta_1 * t + \beta_2 p_{t-10} + \beta_3 \Delta i_{t-10} + \beta_4 bs_{t-10} + \beta_5 be_{t-10} + \beta_6 \mathrm{q} + \beta_7 \Delta gp_{t-10} \tag{2}$$

Thereby Y_t denotes the number of cows in herds with 100 or more cows. In order to adjust the temporal trend in our dependent variable, we have included the dummy variable t in the model. Due to the trend variable we can interpret the other explanatory variables to be as estimated effects with no time trend and since we avoid the estimation of apparent contexts. The lagged milk price is denoted by p_{t-10}, whereas Δi_{t-10} represents the differenced, lagged interest rates for long time loans. bs_{t-10} denotes the lagged current business situation while the lagged business expectations are represented by be_{t-10}. The dummy for the milk quota is indexed by q. The differenced grain price index is represented by Δgp_{t-10}.

5 Results and Discussion

The results in Table 3 show that the milk quota had a clearly, highly significant negative effect on the development of the number of cows in herds of 100 and more cows. In contrast, the development of the interest rate for long-term loans had the greatest posi-

tive effect on the development of the large herds in German dairy farming. The milk price also had a significant positive effect on the development of the large herds in Germany, albeit at a much lower level. Business expectations have a slightly lower but highly significant impact on development the development of this herd size. On the other hand, we found no significant influence of the assessment of the current business situation, nor of the opportunity costs in the form of the grain price index.

Table 3. Results of the Newey-West regression model

Herd size	Coefficient		Robust Std. Err.	P>t
T	10.413	***	0.744	0.000
Millk price quarterly	8.025	*	4.772	0.096
Δ Interests Credits	-113.549	**	51.853	0.031
Business situation	-1.112		1.324	0.403
Business expectations	5.331	***	1.343	0.000
Dummy milk quota	-336.419	***	57.438	0.000
Δ Grain price	-1.224		1.398	0.383
Constant	778.870	***	188.186	0.000

*p<0.1 **p<0.05 ***p<0.01

Number of obs = 102; F (7, 94) = 169.62; Prob > F = 0.0000

Source: Authors' own calculation

The results of our study largely coincide with those from the literature as we find both the negative effects of the milk quota on the development towards larger and thus mostly more competitive units (MacDonald *et al.*, 2007; Kleinhanß *et al.*, 2010). Looking at Figure 1, it can be seen that the removal of the regional link as well as the increase in quota volumes from 2006 onwards favored the development of the large herds. The studies by Kleinhanß *et al.* (2010) and Lassen, *et al.* (2008) are supporting this. Due to the strong increase in the number of animals in the herd size with more than 100 cows and the modest increase in quota amounts in the wake of the abolishment, we can assume that the liberalization of the quota trade has a greater impact on this development. Due to this, the production factor quota became mobile and could migrate to the farms with the best utilization (Chavas and Magand, 1988).

The drastic fall in interest rates in the wake of the global financial crisis in 2009 has the strongest positive effect on the development of the number of animals in herds with more than 100 animals. Because of the great meaning of labor and capital costs in milk production, farmers used the historical low prices for capital to replace the factor labor by capital due to the lower demand for labor per kg milk in bigger herd sizes (Kreins and Cypris, 2000; MacDonald *et al.*, 2007; Koester and Cramon-Taubadel, 2019). This development was accompanied by a sharp increase in the volume of lending in agriculture and forestry, which is essentially due to the increase in long-term loans. We cannot say what share milk production has on this, but it will not be insignificant. From 2009 to 2018 the number of cows in herds with more than 100 cows increased by 855,734. Assuming the construction of a new cowshed cost on average € 9,087 € (Dorfner and Hofmann, 2018), these development would have been accompanied by investments of € 7.8 billion, whereas the long term loans in this time period increased by € 17.3 billion.

The positive influence of the price also supports the results from the literature. Since the price is a major factor influencing the economic success of the company, the price and the expected future price will influence the decision to expand dairy cattle production (Chavas and Magand, 1988; Elhorst, 1992). The historically high milk prices of over 40 Ct./kg in 2007 for example may have provided a positive incentive for an expansion of milk production.

Since the price has a significant influence on the net return of farms and thus also on long-term investments, it can be assumed that the expected price at the time of the investment decision also has an influence on this and thus on the development of large herds (Chavas and Magand, 1988). Though we do not know which price expectations played a role in the individual investment decisions of farmers, the agricultural outlooks of the OECD and FAO provide a picture of the price prospects at that time. For example, the observed and forecasted cheese price in the 2006 and 2008 publications differed considerably (see Figure 4).

Figure 4. Price forecast for cheese in selected OECD-FAO Agricultural Outlook

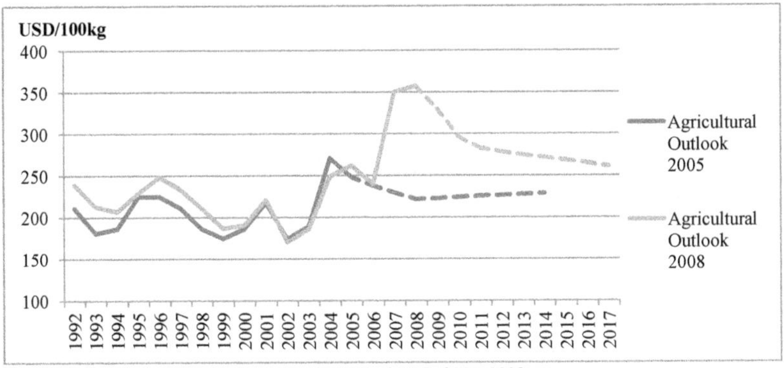

Source: Authors' own depiction based on OECD and FAO, 2005, 2008.

The same applies to the price forecasts for butter, skimmed milk powder and whole milk powder. These increased forecasts are likely to have a positive effect on the development of investment activity and thus herd sizes, which is also underlined by our results as well.

The assessment of the business situation has no significant influence on the number of animals in herds with more than 100 animals and thus on the investments in here, indeed business expectations have. From this, we can conclude that the outlook is more relevant in terms of farmers' investments in dairy production then the current business situation is. Against this background, business expectations seem to be a good complement to the neoclassical variables of production theory to explain investment activity (Elhorst, 1992).

The opportunity costs, considered by the grain price index, have no significant influence on the investments and thus the development of the number of cows in the herds with more than 100 cows. The reason for this result is probably the fact that milk production and thus also investments in Germany have migrated to grassland locations in recent years as a result of the liberalization of quota trading and the expansion of the milk quota, which are characterized by low opportunity costs (Lassen *et al.*, 2008). This is supported by an average share of grassland of 49.1% in highly intensive milk production regions. In 2015 these regions accounted for 67.8% of the growth in milk production. (Meyer and Theuvsen, 2017). Opportunity costs therefore do not play a major role here. However, the higher opportunity costs by alternative use possibilities, e.g., in the form of arable farming, might have contributed to the decline of the animal numbers in particular in the new federal states like Saxony-Anhalt or Thuringia as can be seen in Table 1 (Chavas and Magand, 1988).

6 Outlook and Conclusions

How is the development likely to go on in the future? The expiry of the milk quota at the end of 2015 paved the way for further growth in milk production and thus for the further growth of farms and herd sizes. The interest rates are still historically low and from nowadays perspective are likely to remain low for the foreseeable future according to the latest decision of the European Central Bank (Bundesbank, 2019d; Sueddeutsche Zeitung, 2019). Milk prices are relatively stable at between 30 and 35 cents at the moment, depending on the region and dairy, and are thus at the long-term average level (European Commission, 2019; top agrar, 2019). The expected milk price based on futures for butter and skimmed milk powder shows a slightly positive trend for the nearer future and also the long term projections show a slight increase. Beside this, the projections are also friendly regarding the opportunity costs for dairy production, as, for example the wheat price is projected to decline in real terms until 2028 (ife, 2019; OECD and FAO, 2019). From this point of view the conditions have been created for a further growth of farms and number of cows in large dairy herds.

However, there are some factors that are likely to counteract this. Kleinhanß *et al.* (2010) already pointed out the increasing problem of nutrients in intensive regions in 2010. This problem has become even more acute in the intensive livestock farming regions as a result of the fertilizer ordinance that was tightened in 2017 (Meyer and Theuvsen, 2017). With the renewed tightening that will come into force in 2020, this circumstance will become even worse, especially due to the strong concentration of milk production in these regions in the last years. Due to a constant area stock and decreasing permissible application quantities of animal nutrients, farms are forced to lease additional land, to export surplus nutrients, to outsource individual elements of milk production, such as young cattle rearing, or to reduce the stock which will generally lead to increasing production costs and may slow down or even prevent further growth in milk production (MacDonald *et al.*, 2007; Meyer and Theuvsen 2017; Chamber of Agriculture, 2019). Next to this, other costs, such as construction costs have grown in the past and are likely to grow further in the future. (Dorfner and Hofmann, 2018).

The future development of the milk price is likely to continue to be characterized by high volatility. This increases the liquidity risk of farms, in particular in combination with the sharp rise in loans and the associated debt service. This is likely to have a negative impact on the demand for credits as it depends among other things, on the stability of expected product prices and thus for further growth of farms. The liquidity risk also has a negative effect on the security of the repayment of debt service. As lenders are interested in a secure repayment of the debt service, this could make access to capital more difficult and thus reduce investments (Barry and Robinson, 2001; Schaper *et al.*, 2010; El Benni and Finger, 2013; Koester and Cramon-Taubadel, 2019).

Imperfect information, such as the price volatility mentioned above, also has a negative impact on credit demand. The renewed tightening of the fertilizer regulation in 2020 after its tightening in 2017, for example, falls into this category. Due to the long financing period of generally 20 years for new stables, such changes, which cannot be foreseen at the time of planning, may have a significant impact on the profitability of the project. Incomplete information can also be found, for example, in the discussion about husbandry conditions. Even though dairy farming is still largely unaffected by this, other areas, such as sow farming, show that major changes can occur (Lindena *et al.*, 2018; Koester and Cramon-Taubadel, 2019).

Overall, the trend towards large herds will continue in the future. The abolition of the quota has paved the way for the development of competitive dairy farms in regions that are excellent for milk production, and increasing demands and costs will force farms to further increase efficiency, which they can achieve in larger units through ceteris paribus lower production costs (Kreins and Cypris, 2000; MacDonald *et al.*, 2007). However, due to the increased costs and requirements, combined with volatile prices and high debt servicing due to the high credit level and the resulting liquidity risks, investments and thus the further growth of large stocks are likely to decline overall and concentrate on the most economically efficient farms. The first slight decline in the number of cows in herd size with more than 100 cows since 1996 in 2019 could indicate this.

References

Barry, P.J., Robison, L.J. (2001). Agricultural Finance: Credit, Credit Constraints, And Consequences. In: Gardner, B. and Rausser, G. (Ed.), *Handbook of Agricultural Economics*, Vol. 1, Elsevier Science B.V., Amsterdam: 513–571.

Betzholz, T. (2010). Milcherzeugung vor dem Hintergrund der Kontingentierung. Monthly Statistical Bulletin Baden-Wuerttemberg, 9/2010.

BLE (2019). Milchwirtschaft auf einen Blick nach Kalenderjahren. Federal Office for Agriculture and Food, Bonn.

Bundesbank (2019a). Lending rates banks/long-term fixed rate loans to companies and self-employed from 500,000 to under 5 million euros, effective interest rate/average rate. Deutsche Bundesbank, available at: https://www.bundesbank.de/de/statistiken (accessed 5 May 2019).

Bundesbank (2019b). Effective interest rates for banks DE/new business/loans to non-financial corporations up to 1 million euros, initial fixed interest over 5 years. Deutsche Bundesbank, available at: https://www.bundesbank.de/de/statistiken (accessed 5 May 2019).

Bundesbank (2019c). Diskontsatz of the Deutsche Bundesbank/status at the end of the month. Deutsche Bundesbank, available at: https://www.bundesbank.de/de/statistiken (accessed 5 May 2019).

Bundesbank (2019d). Basiszinssatz in accordance with the Diskontsatz reconciliation law (DÜG)/status at the end of the month. Deutsche Bundesbank, available at https://www.bundesbank.de/de/statistiken (accessed 5 May 2019).

Bundesbank (2019e). Basiszinssatz according to BGB/status at the end of the month. Deutsche Bundesbank, available at: https://www.bundesbank.de/de/statistiken (accessed 5 May 2019).

Bundesbank (2019f). Lending to agriculture and forestry, fisheries and fish farming / long-term/all banking groups. Deutsche Bundesbank, available at: https://www.bundesbank.de/de/statistiken (accessed 5 May 2019).

Bundesbank (2019g). Lending to agriculture and forestry, fisheries and fish farming/mid-term/all banking groups. Deutsche Bundesbank, available at: https://www.bundesbank.de/de/statistiken (5 May 2019).

Bundesbank (2019h). Lending to agriculture and forestry, fisheries and fish farming/short-term/all banking groups. Deutsche Bundesbank, available at: https://www.bundesbank.de/de/statistiken (5 May 2019).

Brümmer, B., Loy, J.-P. (2004). Einführung der Milchquotenbörse in Österreich. Expert opinion on behalf of the Federal Ministry of Agriculture, Forestry, Environment and Water Management, Vienna.

Chavas, J.P., Magand, G. (1988). A dynamic analysis of the size distribution of firms: The case of the US dairy industry. *Agribusiness* 4(4): 315–329.

Chamber of Agriculture (2019). Nährstoffbericht für Niedersachsen 2017/2018. Chamber of Agriculture of Lower Saxony, Oldenburg.

Council of the European Union (1998). Council Regulation (EC) No. 2866/98 of 31 December 1998 on the conversion rates between the euro and the currencies of the Member States adopting the euro. Council of the European Union, Brussels.

Croux, C., Dhaene, G., Hoorelbeke, D. (2003). Robust Standard Errors for Robust Estimators. Katholieke Univeriteit Leuven, Discussion Paper Series 3.16.

DBV (2019). Situationsbericht 2018/2019. Deutscher Bauernverband e.V., Berlin.

Destatis (2011–2019a). Viehbestand und tierische Erzeugung 2011–2019. Fachserie 3, Reihe 4. Federal Office for Statistics, available at: https://www.destatis.de/DE/Themen/Branchen-Unternehmen/Landwirtschaft-Forstwirtschaft-Fischerei/Tiere-Tierische-Erzeugung/_inhalt.html (accessed 1 June 2019).

Destatis (2019b). Grain price index from 1991–2019. Genesis-Online database, Federal Office for Statistics, available at https://www-genesis.destatis.de/genesis/online/data (accessed 1 June 2019).

Dialer, D., Lichtenberger, E., Neisser, H. (2010). *Das Europäische Parlament – Institutionen, Visionen und Wirklichkeit.* Innsbruck University Press, Innsbruck.

Dorfner, G., Hofmann, G. (2019). Baukosten von Michviehställen. Bavarian State Research Center for Agriculture, Freising.

El Benni, N, Finger, R. (2013). Gross revenue risk in Swiss dairy farming. *Journal of Dairy Science* 96(2): 936–948.

Elhorst, J.P. (1992). The estimation of investment equations at the farm level. *European Review of Agricultural Economics* 20: 167–182.

European Commission (2019). Historical EU price series of cow's raw milk, Milk Market Observatory, available at: https://ec.europa.eu/info/food-farming-fisheries/farming/facts-and-figures/markets/overviews/market-observatories/milk (accessed 30 August 2019).

Eurostat (2019). Cows's milk collection and products – monthly data. Eurostat database, available at: https://ec.europa.eu/eurostat/de/data/database (accessed18 July 2019).

FAOSTAT (2018). Production quantity of milk, whole fresh cow, 1991 to 2016. Food and Agriculture Organization of the United Nations, available at: http://www.fao.org/faostat/en/ (accessed 20 May 2018).

Höller, T. (2015). Die Ära der Milchquotenregelung geht zu Ende – Rückschau auf ein tiefgreifendes Instrument. Bavarian State Research Center for Agriculture, Freising.

Hörl, M. and Hess, S. (2017). The Export Competitiveness of the European Dairy Industry. Contribution at the XV Congress of the European Association of Agricultural Economists, 28 August – 1 September, Parma, Italy.

ife (2019). Kieler Börsenmilchwert vom 13.09.2019. Institute for Food Industry, Kiel.

ifo-Institut (2018). Calculating the ifo Business Climate. Ifo-Institute, available at: https://www.ifo.de/umfrage/ifo-geschaeftsklimaindex (accessed 20 December 2018).

ifo-Institut (2019). Business situation and business expactation in the dairy sector. Ifo-Institute, Munich.

Jacobsen, B.H. (1996). Farmers' machinery investments. In: Beers, G., Huirne, R.B.M., Pruis, H.C. (Ed.), *Farmers in small-scale and large-scale farming in a new perspective*, Agricultural Economics Research Institute (LEI-DLO): 261–275.

Kirner, L. (2009). Auswirkung der vollständigen Implementierung des Health-Check auf landwirtschaftliche Betriebe. *Ländlicher Raum*, online journal of the Austrian ministry for Agriculture, Forestry, Environment and Water Conservancy, available at: https://www.bmnt.gv.at/land/laendl_entwicklung/zukunftsraum_land_masterplan/Online-Fachzeitschrift-Laendlicher-Raum/archiv/2009/kirner.html (accessed 20 August 2019).

Kleinhanß, W., Offermann, F., Ehrmann, M. (2010). Evaluation of the Impact of Milk quota – Case Study Germany. Thünen Working Papers, Institute of Farm Economics, Thünen-Institute, Braunschweig.

Koester, U., Cramon-Taubadel, S. (2019). Besonderheiten der landwirtschaftlichen Kreditmärkte. Leibnitz Institute of Agricultural Development in Transition Economics, IAMO Discussion Papers, No. 185, Halle.

Kreins, P., Cypris, C. (2000). Entwicklung der regionalen Wettbewerbsfähigkeit im Bereich der Milchproduktion und Folgen für die Landnutzung. *Schriften der Gesellschaft für Wirtschafts- und Sozialwissenschaften des Landbaus e.V.* 36: 29–36.

Kreß, B. (2005). Die EU-Agrarreform vom 26.6.2003 und die Konsequenzen für die Milchwirtschaft – Expertenbefragung in Dänemark, Österreich und Deutschland. Dissertation, Technical University Munich, Munich.

Lassen, B., Isermeyer, F., Friedrich, C. (2008). Milchproduktion im Übergang – eine Analyse von regionalen Potenzialen und Gestaltungsspielräumen. Thünen Working Papers, Institute of Farm Economics, Thünen-Institute, Braunschweig.

Lindena, T., Tergast, H., Ellßel Raphaela, Hansen, H. (2018). Steckbriefe zur Tierhaltung. Institute of Farm Economics, Thünen-Institute, Braunschweig.

Lutter, M. (2009). Strukturwandel in der Milchviehhaltung: Ergebnisse einer regional differenzierten Befragung. Thünen Working Papers, Institute of Farm Economics, Thünen-Institute, Braunschweig.

MacDonald, J.M., O'Donoghue, J., McBridge, W., Nehring, R.F., Sandretto, C.L., Mosheim, R. (2007). Profits, Costs, and the Changing Structure of Dairy Farming. United States Department of Agriculture, Economic Research Service, Report No. 47.

Meyer, J., Theuvsen, L. (2017). Intensive Dairy Farming in Northern Germany: Development and Impact of the New Fertilizer Act. In: Proceedings of Agrarian Perspectives XXVI, Competitiveness of European Agriculture and Food Sectors, Proceedings 26th International Scientific Conference, 13–15 September, Prague: 219–225.

OECD, FAO (2005). OECD-FAO Agricultural Outlook 2005–2014. OECD Publishing, Paris.

OECD, FAO (2008). OECD-FAO Agricultural Outlook 2008–2017. OECD Publishing, Paris.

OECD, FAO (2019). OECD-FAO Agricultural Outlook 2019–2028. OECD Publishing, Paris.

Offermann, F., Kleinhanß, W., Manegold, D. (2002). Ausstieg aus dem Milchquotensystem: wie und mit welchen Folgen? Landbauforschung Völkenrode, special issue 242: 91–96.

Rottmann, H., Auer, B. (2010). Ein Überblick über die lineare Regression. *WiSt – Wirtschaftswissenschaftliches Studium* 39(11): 548–554.

Rougoor, C.W., Mandersloot, F., Hanekamp, W.J.H., Hurine, R.B.M., Dijkhuizen, A.A. (1996). Data analysis to quantify the impact on management on dairy farm performance. In: Beers, G., Huirne, R.B.M., Pruis, H.C. (Ed.), *Farmers in small-scale and large-scale farming in a new perspective*, Agricultural Economics Research Institute (LEI-DLO): 61–69.

Sauer, J., Lactaz-Lohmann, U. (2015). Investment, technical change and efficiency: empirical evidence from German dairy production. *European Review of Agricultural Economics* 42(1): 151–175.

Schaper, C., Lassen, B., Theuvsen, L. (2010). Risk management in milk production: A study in five European countries. *Acta Agriculturae Scandinavica, Section C – Food Economics* 7(2–4): 56–68.

Stigler, G. (1958). The Economics of Scale. *Journal of Law and Economics* 1: 54–71.

Sueddeutsche Zeitung (2019). EZB drückt Zinsen noch niedriger. Süddeutsche Zeitung online, available at: https://www.sueddeutsche.de/wirtschaft/ezb-zinsentscheidung-strafzins-1.4598022 (accessed 16 September 2019).

Tangermann, S., Cramon-Taubadel, S. (2013). Agricultural policy in the European Union: An overview. Working paper No. 1302, Department of Rural and Agricultural Development, Georg-August-University Göttingen, Göttingen.

Top agrar (2019). Milchpreisbarometer. Online platform of top agrar, available at: https://www.topagrar.com/rind/milchpreisbarometer/nord/ (accessed 1 October 2019).

Vasaveda, U., Chambers, R.G. (1986). Investments in U.S. Agriculture. *American Journal of Agricultural Economics* 68(4): 950–960.

Weiss, C.R. (1999). Farm Growth and Survival: Econometric Evidence For Individual Farms in Upper Austria. *American Journal of Agricultural Economics* 81: 103–116.

ZMB 2012. ZMB Jahrbuch Milch 2012. Zentrale Milchmarkt Berichterstattungs GmbH, Berlin

Appendix

Table 1. Results from the Augmented Dickey-Fuller-Test

Variable	Test Statistic	1% Critical Value	5% Critical Value	10% Critical Value
Herd size	-1.303	-4.04	-3.45	-3.15
Milk price quarterly	-3.112	-3.51	-2.89	-2.58
Δ Interests credits	-5.711	-3.51	-2.89	-2.58
Business situation	-2.963	-3.51	-2.89	-2.58
Business expactation	-5.525	-3.51	-2.89	-2.58
Δ Grain price	-6.365	-3.51	-2.89	-2.58

Source: Authors' own calculation

Table 2. VIF-statistics

Variable	VIF	1/VIF
Business situation	2.84	0.3522
Business expectations	2.11	0.4728
Milk price	1.52	0.6570
Time trend variable	1.62	0.6166
Milk quota	1.63	0.6117
Δ Interests Credits	1.10	0.9101
Δ Grain price	1.11	0.8969
Mean VIF	**1.71**	

Source: Authors' own calculation

IV Assessing the International Competitiveness of the German Dairy Industry by Analyzing Foreign Trade

Authors: Johannes Meyer, Sebastian Lakner, Jan-Henning Feil, Christian Schaper

In preparation for submission to: Agricultural Economics (AGRICECON)

Abstract

Regional imbalances of production and demand and a further liberalization of markets leads to an increasing internationalization in the dairy industry and offers new markets to producers. Successful competing in international markets requires international competitiveness. There is no uniform, defined method for determining international competitiveness. Rather, the assessment of international competitiveness comes from a variety of perspectives. One of them is an assessment based on foreign trade-based indicators. This paper examines the trade structure and international competitiveness of the German dairy industry and nine other competitors from 2001–2017 for six product groups using foreign trade data-based metrics. While New Zealand and Belarus show a high to very high degree of international competitiveness across all product groups, the German dairy industry and some other countries have a different profile depending on the product groups considered. Overall, the international competitiveness of the German dairy industry has slightly increased over the past 17 years.

1 Introduction

The dairy industry has been facing a process of rapid internationalization due to trade liberalization and regional imbalances on the world milk market (Guillouzo and Ruffio, 2005; Heyder *et al.*, 2011; Vitaliano, 2016). As a consequence, international trade flows between net exporters, such as New Zealand, the United States and the European Union, and net importers, such as China, the Middle East and Africa, have been increasing. At the organizational level, this development has resulted in growing export activities on the part of dairy companies located in net export regions (Trademap, 2018).

This situation can be illustrated by a review of the German dairy sector, which has faced constant increases in milk production over the course of the last decade. During the same time period, domestic demand has remained more or less stagnant, leading to a growing milk surplus. This surplus had to be exported, and German dairies were increasingly forced to look for market opportunities abroad (Heyder *et al.*, 2011).

As a consequence, German dairy export rates increased from 27.8% in 2008 to 33.4% in 2017, peaking in 2014 with exports accounting for 34.2% of total industry turnover (Destatis, 2019a, 201b). After 2014, the milk exports slightly declined.

A number of international circumstances have affected the overall trade balance of Germany and the EU respectively (EU Commission, 2016). We can observe a number of developments in the Germany milk industry, which can lead to a positive change in competitiveness:

- **Increased milk production in Germany** in view of the quota abolishment since 2009 due to investments into stables and production capacities which was also supported by the low interest rates for loan capital following the financial crisis in 2009 (Eurostat 2019; EU Commission 2016). Even decreasing milk price levels after 2014 (AMI, 2015, 2018), and an increased structural change after 2010, leading to increased herd sizes suggest an increased level of competitiveness.
- Other **production factors** like land and labor rather describe limits in competitiveness: The factor land is limited, and we can observe increasing land prices. Labor costs are amongst the highest within the EU, which might also be regarded as challenge especially against competitors from Eastern EU countries or from Ireland, where labor costs are lower.

- In the dairy business, we could observe a **concentration process**, estimated e.g. with a GINI-coefficient increasing from 0.61 (in 2008) to 0.66 (2016) (own calculation based on Destatis, 2008–16), suggesting larger processing units. As an outcome of this concentration process, the DMK, founded in 2010 based on fusion of Hansano and Humana, appeared as the first German dairy as global player: DMK nowadays is #13 of the world's largest dairies (Rabobank, 2018). On the trade side, we can also observe fundamental changes in **trade conditions**, where policies and international conflicts have changed the trade environment, in which the trade of milk is taking place.
- The import **ban of agricultural products by Russia** in 2014 due to the conflict around the Russian occupation of Crimea and the war in Eastern Ukraine also affected the export opportunities of milk producer in the EU and US, Norway, Canada and Australia. This ban has reduced agri-food imports from the EU to Russia from € 11.8 bn. to € 6 bn. Though also other agri-food sectors have been affected by the Russian export ban, the German dairy sector suffered most by price pressure due to unrealized exports (EU Commission, 2019; Fedoseeva, 2016).
- **Declining import demand by China** in 2015 reduced the milk price in the same direction (EU Commission, 2016).
- **Worldwide growing demand** for milk and dairy products, but strong inventory build-up in 2014 and 2015 due to a higher production surplus compared to previous years (ZMB, 2016).
- The **Brexit** vote from 29th of March 2016 with a scheduled exit of the United Kingdom 29th of March 2019. As we know now, this exit is delayed to October 2019. Therefore, it is yet not possible to estimate the impacts of the Brexit, which will heavily depend on the Brexit-option chosen by the UK government. In case of a "No-Deal Brexit" the trade with dairy products which count to the simple processed products, is predicted to decrease by 95%. Thereby Non-Tariff-Measures will play an important role, as trade without these is predicted to decrease just about the half in case of a "No-Deal Brexit". The effect of less imports from the UK and therefore higher demand on the domestic markets for the European dairy products will not compensate the losses in value added (Bellora *et al.*, 2017; Brümmer, 2019).

The relevance of exports is even greater considering the physical amounts being exported. In 2016, German dairies processed 33.8 m. tons of milk, including 2.5 m. tons of raw milk imports from neighboring countries. Of this, 16.6 m. tons of milk equivalent were exported as cheese, whole or skimmed milk powder and other products, which corresponds to 49.1% of processed milk (MIV, 2019). Internationalization has therefore become the foremost driver of industry development (Theuvsen *et al.*, 2010).

Some studies already worked on the international competitiveness of the German dairy industry based on trade data-based measures. Hörl and Hess, 2017 analyzed the export competitiveness of the German dairy industry compared to its European competitors. Drescher and Maurer (1999) analyzed the international competitiveness of the German dairy industry for different products groups in the period from 1983 to 1993. Weindlmeier (1999) analyzed the international competitiveness of different product groups of the German dairy industry in the period from 1988 to 1997. Beyond the dairy industry in Germany, a number of studies have analyzed export competitiveness of the agri-food sector within the European Union (Poppe *et al.,* 2008; Bojnec & Fertő, 2007) or for Slovakia (Simo *et al.,* 2016), using trade-flow data.

On the one hand, some of the early studies using data from the 1990s were not yet updated; on the other hand, important information is lost through the method of aggregating[1] different product groups.

What they all have in common is that they compare the international competitiveness of the German dairy industry with their European competitors and not with major competitors outside the European Union. In the course of the changed conditions, such as the abolition of the milk quota, along with the increasing liberalization of global milk markets, the German dairy industry faces new challenges and new competitors competing for global sales markets. In addition, when assessing international competitiveness, we must consider who we are struggling to gain market share in which markets and which products.

Based on the outlined developments, we investigate, whether we can observe an increased export competitiveness of the German dairy sector. Based on trade data, we

[1] According to Vollrath 1991, there can be a loss of information by aggregation. Though a country might have a competitive disadvantage in a sector, here expressed as the aggregated commodities of milk products, it might have a competitive advantage for specific products.

model index-numbers describing Germany's position in international trade with milk and dairy products, which show how the international competitiveness of the German dairy sector has developed over time. We also take into account the phasing out of the milk quota, which began in 2009 with a gradual increase in quota volumes by looking at two periods, the first from 2001 to 2009, the second from 2010 to 2017. We analyze the structure of the German dairy exports and determine its competitors for different products on the corresponding markets. Thereby we will also refer to important political changes which have happened over the observed time. On that basis we assess the international competitiveness of the German dairy industry against its major competitors for six different product categories of dairy products based on foreign trade data over a 17-year time period.

2 Theoretical Background

The economic literature has no consensus on the definition of international competitiveness yet (Gries and Hentschel, 1994; Feurer and Chaharbaghi, 1994; Trabold, 1995; Baade, 2007). Any definition must be based on the notion of competition, which generally refers to the pursuit of a goal by at least two actors, with the higher degree of goal attainment of one usually entailing a lower degree of goal attainment by the other. At the same time, competition is by no means limited to the economy (Winter and Würmann, 2012). The essential feature of competition, therefore, is the conflict of interests of two or more individuals, groups or institutions (e.g., companies, industries, states) that express their desire to be more successful than others (Reiljan *et al.*, 2000). Competitiveness is, therefore, the ability to achieve a higher degree of target achievement than competitors in the competitive show characterized by competition (Man *et al.*, 2002; Nelson, 1992; Gries and Hentschel, 1994). When analyzing international competitiveness, it is first necessary to clarify the question of the level at which competitiveness is to be investigated. In general, a distinction is made between national, industry and company levels (Buckley *et al.*, 1988). In this work, the aggregation level is determined by the analysis of the German dairy industry. International competitiveness is therefore analyzed at industry level. The aggregation level is important not only because of the choice of analysis methods. At the national level, despite numerous publications in the literature, there is no uniform definition of the competitiveness of countries or economies (Trabold, 1995). Although the definition of international competitiveness of companies is not clearly defined, these definitions in the literature are more or less in the same direction.

Based on the prevailing literature definitions, we consider the international competitiveness of the German dairy industry based on Utzig (1987), Martin *et al.* (1991) and Weindlmaier (1999) as "the ability to sell their products on foreign markets for sustained, full cost covering prices, while gaining market share or at least maintaining its market share". We refer to the German dairy industry as the sum of all the companies included in the statistics, which is why the industry has the same criteria for assessing international competitiveness as it does for an individual company. There are several methods available to analyze competitiveness at the industry level. Examples of methods for assessing international competitiveness are the analysis of foreign direct investments, the analysis of trade and income related data and the evaluation of the annual reports of firms in terms of business metrics. In addition to these quantitative analyzes, there are also qualitative approaches, such as the Diamond Model by Michael Porter (Frohberg and Hartmann, 1997; Weindlmaier, 1999). Both the definition of the international competitiveness of the German dairy industry and the multitude of measurement approaches available make it clear that a final assessment of international competitiveness can only deliver a holistic result if a large number of perspectives are considered.

This work will analyze the international competitiveness of the German dairy industry on the basis of its foreign trade with its major competitors. Trade data-based metrics are based on the idea that companies or industries are competitive if they are able to offer their products abroad without subsidizing government support and at competitive prices (Weindlmaier, 1999). Common to these indicators is that their results include information on supply and demand conditions for production factors and goods (Vollrath, 1991).

A widely used method in the literature for determining international competitiveness is Balassa's Revealed Comparative Advantage Index (RCA). It is based on the idea that the observed trade patterns of individual goods reflect relative costs and differences in non-price factors. Thus, the international competitiveness of individual countries based on the trade structure of different goods compared to other countries can be determined with the help of the RCA (Balassa, 1962; Vollrath, 1991).

However, the RCA is only of limited suitability for the cross-national comparison of different sectors regarding their international competitiveness, since the size of the country has an influence on value if the countries have different levels of total exports. Rather, the RCA is better suited to identifying industries within a country with comparative cost advantages and disadvantages (Weindlmaier; 1999; Trail and Pitts, 1998). The RCA measures only success or failure, but it can give hints where to investigate relevant industries further, on causes of successes or failures (Trail and Pitts, 1998). Moreover, the RCA index according to Balassa does not take into account intra-industrial trade in any way. Intra-industrial trade refers to the simultaneous import and export of products and goods belonging to the same sector (Vollrath, 1991).

Despite these limitations, the RCA is a widely used method in the literature to assess the comparative cost advantage of whole nations and individual industries. Balassa itself used the RCA to analyze the comparative cost advantage of various industrialized countries in 74 industrial goods. Over time, RCA has been constantly evolving and changed by various scientists to address problems such as to consider the lack of regard for intra-industrial trade (Vollrath, 1991). This paper will use the Relative Trade Advantage (RTA), Relative Export Advantage (RXA) and Relative Import Penetration Index (RMP) indices presented by Vollrath (1991), which are also derived from the RCA, to analyze the international competitiveness of different states for various dairy products but bypass major disadvantages of RCA. Though originally used for the analysis of manufacturing industries in developed countries, the RCA or derived indexes, such as the RTA, RXA or RMP, have been used in several studies to assess the international competitiveness of individual agricultural industries and products.

Vollrath (1987, 1989) analyzed international competitiveness in agricultural commodities between the United States and its competitors over time, using RTA, RXA, RMP and market shares. Weindlmeier (1999) also used RTA, RXA and RMP for analyzing the international competitiveness of the German food industry. He differentiated between different milk, meat and plant origin products and considered evolution over time by analyzing two different time periods. Drescher and Maurer (1999) analyzed the international competitiveness of the European dairy industry using the RCA related Revealed Comparative Advantage Export Indicator (XRCA) and Net Export Indicator (NXRCA). Fertö and Hubbard (2003) also used RTA, RXA and RMP for analyzing international competitiveness in different Hungarian agricultural products, such as wheat, pork and milk exports to the European Union. They also refer to advantages by comparing the RCA mentioned above. They analyzed international competitiveness over a time period of seven years. Serina and Civan (2008) analyzed the RCA of Turkish olive oil, tomato and fruit juice involving major competitors outside the European Union and the whole European Union over a time period of ten years. Ishchukova and Smutka (2013) analyzed the international competitiveness of Russian agriculture with the help of Balssa's RCA and Vollrat's RTA. They looked at 19 product groups such as milk and milk products over a period of ten years. Bojnec & Fertö (2011) use RCA with a Markov chain transition matrix to analyze the international competitiveness of the European dairy industry. They find the Germany dairy sector as competitive for the extra-EU trade, but to be less competitive within the EU-trade (Bojnec & Fertö, 2011).

The use of Balassa's RCA and Vollrath's RTA, RXA and RMP indices is therefore widely employed in the literature of agricultural field studies. All the works involved, have each analyzed the international competitiveness of the particular sectors or products over time. In this way, far better conclusions can be drawn with regard to the development of individual international states' competitiveness than as a result of a specific observation at a given time. Many of the studies considered analyzed international competitiveness using the indices mentioned at the sector level. This is possible, but the approach can generalize and burr the results through the aggregation of individual products into one sector (Vollrath, 1991). For example, indices may point to poor international competitiveness for a sector, while individual products may be more competitive. By themselves, however, individual figures are not very meaningful. Rather, they have to be compared with corresponding values from the most important competitors in order to make a statement about the international competitiveness (Hörl and Hess, 2017). Some of the studies mentioned above, simply show values for the country observed and its sectors. As a result, it is difficult to formulate a substantiated international competitiveness statement, as any comparison with competitors is missing.

3 Data and Method

The analysis is based on data from the trade data platform Trade Map of the International Trade Center (ITC). Data over a period of 17 years from 2001 to 2017 were collected and analyzed for 10 countries and captured for the subgroups 0401 to 0406 of product category 04.

The countries included in the analysis were Germany, United States of America, New Zealand, Netherlands, France, Ireland, Poland, Czech Republic, Argentina and Belarus. All these countries play an important role on the international dairy export market and show positive trade balances in dairy trade. In 2017 these countries represented 60.7% of the world wide dairy exports. The six product groups supported are based on the classification of the Harmonized System, an international nomenclature for the classification of goods in the World Customs Organization (WCO). We opted for the 4 digit level because we believe it is a good compromise between the necessary aggregation in order not to get lost in unnecessary details and the necessary detail, especially in comparison with previous studies. We divided the results in two periods, the first from 2001 to 2009, the second from 2010 to 2017. We chose this division due to unique increase of 2% of the milk quota in the 2008/2009 season and the following incremental increase in quota amount of 1% per year, starting in the 2009/10 season in preparation of the quota abolishment in the European Countries under consideration in 2015 (Dialer *et al.*, 2010; Kirner 2009). All value information such as export and import values were calculated in Euros. The following data were used to calculate the key figures: export and import values of the corresponding product group of the individual countries, worldwide export and import values of the individual product groups, total export and import over all products of the individual countries, as well as the world-wide total exports and imports over all goods and all countries.

The following product groups have been analyzed:

Product group 1: Milk and cream, not concentrated nor containing added sugar or other sweetening matter

Product group 2: Milk and cream, concentrated or containing added sugar or other sweetening matter

Product group 3: Buttermilk, curdled milk and cream, yogurt, kefir and other fermented or acidified milks and creams, whether or not concentrated or flavored or containing added sugar or other sweetening matter, fruits, nuts or cocoa

Product group 4: Whey, whether or not concentrated or containing added sugar or other sweetening matter; products consisting of natural milk constituents, whether or not containing added sugar or other sweetening matter

Product group 5: Butter, including dehydrated butter and ghee, other fats and oils derived from milk; dairy spreads

Product group 6: Cheese and curd

In the beginning we will analyze the structure of the German dairy export market and figure out which are important markets for which products and which are the corresponding competitors on these markets. In order to assess the international competitiveness of the dairy industries of the considered countries, the following key figures are used:

A common, easy-to-calculate indicator is the export share of a country for a given commodity on the world market (XS) (Buckley *et al.*, 1988).

$$\mathrm{XS} = X_j^i / X_j \tag{1}$$

Here X_j^i stands for the exports of the good considered j from country i. X_j denotes the total exports from all countries of the good j. Any higher export share can be interpreted as an indication of enhanced international competitiveness. In order to depict international competitiveness over time, export shares of the first and last observation are compared. A positive development is an indication of increasing international competitiveness, while falling shares point to competitive disadvantages (Poppe *et al.*, 2007).

As noted above, RTA, RXA and RMP are based on Balassa's RCA index. This is calculated as follows:

$$RCA_j^i = \{X_j^i / X_j\} / \{X^i / X\} \tag{2}$$

X_j^i denotes the exports of the commodity j from the country i, X_j the exports of the commodity j from all countries (Yu *et al.*, 2008). RCA values> 1 indicate a competitive advantage of country i for good j, while values <1 are indicative of competitive disadvantages (Weindlmaier, 1999).

To assess international competitiveness for various product groups, this paper uses the Relative Trade Advantage (RTA). It is calculated from the difference between Relative Export Advantage (RXA) and Relative Import Advantage (RMP).

$$RXA_j^i = \{X_j^i / \textstyle\sum_{l,l\neq i} X_j^l\} / \{\textstyle\sum_{k,k\neq j} X_k^i / \textstyle\sum_{k,k\neq j} \textstyle\sum_{l,l\neq i} X_k^l\} \tag{3}$$

The RXA Index is defined as the ratio of the export share of a given good (j) of a country (i) on the world market ($X_j^i / \sum_{l,l \neq i} X_j^i$) to the share of the export share of all exported goods (k) of the country under consideration in the total export of all countries (l) ($\sum_{k,k \neq j} X_k^i / \sum_{k,k \neq j} \sum_{l,l \neq i} X_k^l$). In contrast to the calculation of the RCA, here at the summation of all countries the considered country is not taken into account (l, l $\neq$ i). Furthermore, when the exports of all goods of all countries are summed, the exports of the goods considered are not taken into account (k, k $\neq$ j). Preventing numerator and denominator double counting avoids index distortions that can occur when a country has large proportions of international trade, or when the commodity under consideration has a significant share in world trade. Positive RXA values imply a relative export advantage, negative values accordingly indicate a relative export disadvantage of the considered country considered good (Frohberg and Hartmann, 1997)

$$RMP_j^i = \{M_j^i / \sum_{l,l \neq i} M_j^i\} / \{\sum_{k,k \neq j} M_k^i / \sum_{k,k \neq j} \sum_{l,l \neq i} M_k^l\} \quad (4)$$

The RMP Index is calculated analogously to the RXA Index but referring to imports (M), whereby the interpretation is exactly the opposite. Positive values indicate a relative import disadvantage in the RMP as they reflect a high proportion of imports compared to other sectors of the country. Negative values on the other hand, reflect a relative import advantage (Frohberg and Hartmann, 1997; Weindlmaier, 1999).

$$RTA_j^i = \{X_j^i / \sum_{l,l \neq i} X_j^i\} / \{\sum_{k,k \neq j} X_k^i / \sum_{k,k \neq j} \sum_{l,l \neq i} X_k^l\} - \{M_j^i / \sum_{l,l \neq i} M_j^i\} / \{\sum_{k,k \neq j} M_k^i / k, k \neq jl, l \neq iMkl \quad (5)$$

The RTA Index for a considered country and a considered good is determined by the relative export advantage and relative import advantage. In the case of RTA, positive values indicate a competitive advantage in the good under consideration, while negative values indicate a competitive disadvantage. Positive values result from relative export advantage exceeding relative import advantage, negative values from relative import disadvantage surpassing relative export advantage (Scott and Vollrath, 1991; Frohberg and Hartmann, 1997).

In contrast to the RCA calculation there is a distinction between a specific commodity and all other commodities, as well between a specific country and the rest of the world, as proven in the calculation of the denominator of the RXA and RMP (Vollrath, 1991). An advantage of this measure is that it takes into account intra-industrial trade by accounting for exports and imports. Intra-industrial trade occurs when a country exports as well as imports goods from the same industrial sector (Marvel and Ray, 1987). It also avoids distortions in the assessment of international competitiveness resulting from, for example, the country being considered as a transit country. The sole consideration of the export shares, the RCA or the RXA would easily lead to misjudging the international competitiveness in such a country (Frohberg and Hartmann, 1997). As a result, the RTA index shows how a particular good from a country competes domestically as well as globally within the context of import demand. Due to the fact that domestic and foreign demand is embodied in this index, extent is described as competition for domestic resources between producers of a particular product within that country. Furthermore, it shows the ability of the country to compete with that particular product in global markets (Vollrath, 1992).

4 Results

4.1 Structure of the German Dairy Export

Germany exported milk and milk products with a total value of € 8.6 bn. which corresponds to 12.4% of the world's total exports of milk and milk products of € 69 bn. in 2017. 83.3% of these exports went to Member States of the European Union (see Table 1). From 2001 to 2017 the share of exports to Member States of the European Union decreased slightly from 87.2% by 3.9 percentage points. Also, the export market share on world dairy exports decreased slightly from 15.7% in 2001 by 3.3 percentage points to 12.4% in 2017.

As can be seen in Table 1 there are great differences regarding the export structure of the different product groups considered. With an export value of € 3.9 bn. in 2017 cheese and curd have by far the highest export values in German dairy exports, followed by concentrated milk and cream (€ 1.3 bn.) and not concentrated milk and cream (€ 1.3 bn.). Over the observed time period from 2001 to 2017 Germany lost export market share on world dairy markets in every product group except in butter which increased from 5.4% in 2001 by 2.8 percentage points to 8.2% in 2017.

Table 1. Structure and performance of German dairy exports

	Export value (bn. €)	**XS: Export share** (%)	**EU exports** (%)	**Trade balance** (m. €)	**Largest import destination outside EU** (Imports m. €) /Export-Share
1. Milk and cream (€ 8.4 bn.)					
2001	1	31.6	98.4	657.8	Mauritania (18.6) 1.4%
2017	1.3	15.7	82.3	-201.4	China (129.7) 9.9%
2. Concentrated milk and cream (€ 16.7 bn.)					
2001	1.1	11.9	76.6	890	Saudi Arabia (51.1) 3.8%
2017	1.3	8	62.9	1,074	China (57.8) 4.3%
3. Buttermilk (€ 4 bn.)					
2001	0.5	27.1	90.4	363.2	Switzerland (8.8) 1.1%
2017	0.8	20.7	88.4	657	China (38.3) 4.6%
4. Whey (€ 4 bn.)					
2001	0.2	16.7	91.4	139.7	Indonesia (13.8) 2.9%
2017	0.5	11.8	78.9	231.5	China (25.4) 5.3%
5. Butter (€ 8.8 bn.)					
2001	0.2	5.4	92.3	-260.3	Iran (6.9) 0.9%
2017	0.7	8.2	91.5	-22.8	USA (9.5) 1.3%
6. Cheese and curd (€ 27.2 bn.)					
2001	1.9	15.2	85.8	-139.3	Japan (51.2) 1.3%
2017	3.9	14.4	88.6	141.3	Switzerland (70.1) 1.8%
All dairy products (€ 69 bn.)					
2001	4.8	15.7	87.2	1,651.0	Switzerland (85.1) 1%
2017	8.6	12.4	83.3	1,879.7	China (260) 3.0%

Source: Authors' own depiction and calculation based on Trademap, 2019.

Regarding trading partners, we can see that the Member States of the European Union are by far the most important importers of German milk and milk products. Indeed, the statistics reveal some differences as well: While the share of exports going to EU member states is highest at butter with a share of 91.5%, followed by cheese and curd (88.6%) and buttermilk, curdled milk, cream and yoghurt (88.4%), the lowest share of intra EU-exports can be seen in concentrated milk products (62.7%), followed by a clear distance from not concentrated milk products (82.3%).

Thereby, the share of exports to EU Member States increased over the observed time period 2001 to 2017 for cheese and curd. In the other products groups considered it decreased, while the strongest decrease can be seen at concentrated milk products. Here the share of EU Member States on the total exports of Germany decreased from 76.6% in 2001 by 13.7 percentage points to 62.9% in 2017. Over all considered product groups, Germany has a trade balance of € 1.9 bn. in 2017 which increased by € 228.7 m., respectively 13.9% from € 1.7 bn. in 2001. The trade balance for the different products reveal some heterogeneity, as presented in Figure 1.

Figure 1. Development of German trade balances for considered product groups

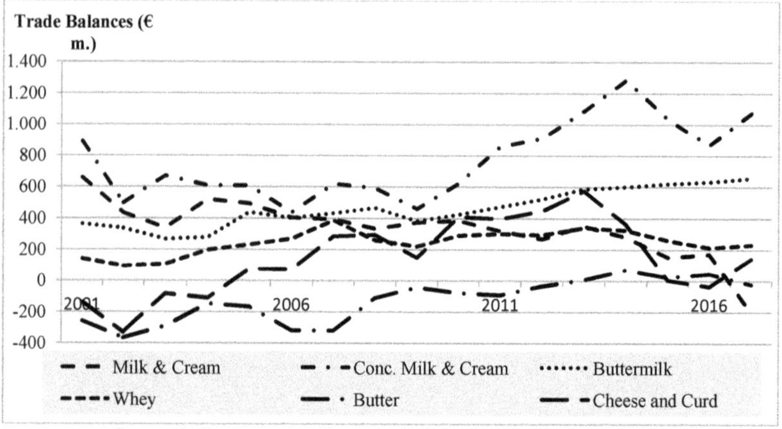

Source: Authors' own depiction and calculation based on Trademap, 2019.

The largest surplus in trade balance origins from concentrated milk products (€ 1.1 bn. in 2017). The exports of these products increased tremendously up to € 1.3 bn.in 2014 since 2010. The second largest trade surplus results from buttermilk, curdled milk, cream and yoghurt, which contributes € 657 m. to the overall trade balance in 2017. From 2001 the trade surplus increased from € 363.2 m. by € 293.8 m. to € 657 m. (2017). In this group we can see a rather constant positive trend within the trade balance. Whey also contributes to the positive trade balance with 231.5 m. in 2017, which also took a positive development since 2001 (+€ 91.8 m.). Contrary, a substantial decrease in trade surplus can be observed for non-concentrated milk and cream and whey, where trade surplus dropped by € 859.2 m. since 2001. Butter and cheese and curd show differences to the other product groups considered: These product groups initially showed negative trade balances in 2001. However, over time, the trade balances in both groups increased and turned positive for cheese and curd with 141.3 m. (€ 2017), whereas despite the positive trend, the trade balance of butter being still negative in 2017 (€ -22.8 m.).

For all considered product groups, China is the most important trading partner outside the EU for the German dairy industry with an import volume of € 260 m. in 2017, respectively 3% of all exports, followed by Switzerland with imports of € 85.1 m., respectively 1% in 2017. Thereby more than the half of Chinas imports in 2017 derived from

non-concentrated milk and cream (€ 129.7 m.). The importance of other countries differs between the different product groups considered (table 1).

4.2 Development of Dairy Trade of Major Competitors

In our study we compare the international competitiveness of the German dairy industry to 9 major competitors on the international dairy markets. These competitors are Argentina, France, Ireland, Netherlands, New Zealand, Poland, Czech Republic, the USA and Belarus.

Analogous to the export values the market shares of New Zealand (12.7%) and the Netherlands (11.3%) are the highest on the world export market for dairy products in 2017 (see Table 2). As at the export value we can see increasing market shares in every considered country except in Argentina. Here the market share decreased minimal over the observed time period from 1.0% in 2001 to 0.7% in 2017. Over all considered countries concentrated milk products and cheese and curd are the most important export products. In case of New Zealand (51.7%) concentrated milk products account for more than the half of their dairy exports in 2017, followed by Argentina with a share of 48.8%. Cheese and curd play the most important role in France and the Netherlands where the exports of cheese and curd account for 50.6% and 45.7% of total dairy exports. Exceptions from this can be seen in Ireland, Poland and the Czech Republic. In Ireland butter is with a share of 39.6% the most important product after cheese and curd (34.7%), in New Zealand butter follows concentrated milk products with a share of 23.9% on total dairy exports. In the Czech Republic non-concentrated milk products are the most important export products with a share of 45.4% on total dairy exports. In Poland the exports of non-concentrated milk products are the second most important with a share of 20.6% on total dairy exports. In both cases the majority of these products, including raw milk is exported to Germany.

Table 2. Structure and performance of competing countries for all product groups

	Exp. (bn. €)	XS %: Export share	Share of Groups on exp. (2017)	Trade balance (m. €)	Largest importers 2016 (Imports m. €) Share on exports
Argentina					
2017	0.6	1.0	2 (48.8%)	485.9	Brazil (199.3) 38.8%
2001	0.3	0.7	6 (34.2%)	280.0	Russia (65.4) 12.7%
France					
2017	5.7	8.7	6 (50.6%)	2377.4	China (328.5) 5.4%
2001	3.9	12.7	2 (12.8%)	1,832.0	USA (181.1) 3%
Ireland					
2017	2.4	3.4	5 (39.6%)	1,174.4	United States (114) 4.8%
2001	1.1	3.4	6 (34.7%)	819.5	Algeria (67.9) 2.9%
Netherlands					
2017	7.8	11.3	6 (45.7%)	4,255.8	China (256.6) 3.3%
2001	3.7	12.1	5 (18.9%)	1,880.3	Hong Kong (133.6) 1.7%
New Zealand					
2017	8.8	12.7	2 (51.7%)	8639.7	China (2,564.1) 29.2%
2001	3.0	9.7	5 (23.9%)	2,953.1	Australia (442.5) 5%
Poland					
2017	2.0	2.9	6 (35.4%)	1,095.10	Algeria (67) 3.5%
2001	0.4	1.3	1 (20.6%)	351.9	China (47.6) 2.5%
Czech Republic					
2017	0.7	1.0	1 (45.4%)	157.9	Bangladesh (14.5) 2%
2001	0.2	0.7	6 (26.7%)	141.3	Lebanon (14) 2.0%
USA					
2017	3.4	5	6 (37.5%)	1,728.6	Mexico (1,031.7) 30%
2001	0.7	2.3	2 (35.4%)	-445.5	China (360) 10.5%
Belarus					
2017	1.9	2.7	6 (37.2%)	1,856.8	Russia (1,714.3) 90.3%
2001	0.17	0.5	2 (22.2%)	151.0	Kazakhstan (63.7) 3.4%

Source: Authors' own depiction and calculation based on Trademap, 2019.

Regarding the trade balance we can see increasing values in all considered countries. The lowest increase in trade balance over the observed time period can be seen in the Czech Republic. From 2001 to 2017 it increased by € 16.6 m., respectively 11.7%. By contrast, the largest increase in the trade balance can be observed in Belarus. In the period from 2001 to 2017, the trade balance increased from € 151 m. by 1,129.7% to € 1.9 bn. Belarus is followed by the United States which showed a negative trade balance of € -445.5 m. in 2001. From this the trade balance increased by 488% to € 1.7 bn. in 2017. New Zealand, which had already a high trade balance in 2001, also more than doubled it by 192.6% to € 8.6 bn. in 2017. Argentina increased its trade balance in the observed time period by 99.6%, though its export market shares slightly decreased over the observed time.

In the group of countries considered, China is the most important buyer referring to individual countries outside of Europe in terms of export value across all product groups considered. China is the most or second most important importer for France, the Netherlands, New Zealand, Poland and the United States. As in the case of Germany, the European Union is the most important export market also for the European competitors which are analyzed in this study as can be seen at the relatively low shares on total exports of the shown third states. Next to these important importers we can find more differences regarding important export destinations of the considered countries. For Argentina Brazil is the most important buyer of its dairy products with an import of € 199.3 m., respectively 38.8% of total exports in 2017. It is followed by Russia which imported 12.7% of Argentina's total dairy exports in 2017. Algeria is the most important importer for Poland and second most important importer of dairy products for Ireland (outside the European Union) with 3.5% and 2.9% of total dairy exports. At 2.0%, respectively, Bangladesh and Lebanon are the main consumers of Czech dairy products outside the European Union. For the United States, Mexico is the largest consumer of dairy products, ahead of China. In 2017, Mexico's share was 30% of total exports while China's imports amounted to 10.5% in 2017. A special situation arises with regard to Belarus. At 90.3% in 2017, almost all exports of milk and dairy products to Russia. Kazakhstan follows in second place with a share of 3.4% in total exports.

4.3 International Competitiveness of Countries in considered Product Groups

The results are ranked by RTA values for the period 2012–2016 in descending order. To underline the development of the countries observed in the different product groups, the Trend-RTA shows the performance of the RTA value from the 2001–2009 period to the 2010–2016 period. Increasing RTA values are indicated by a "+", decreasing values are indicated by a "-".

Non-concentrated milk and cream

Exports of milk and cream had a worldwide export market volume of 8.4 bn. Euros in 2017. The highest RTA values can be seen in New Zealand and Belarus with 12.93 and 11.60 in 2017 (see Table 3). They are followed by the Czech Republic (3.61) and Poland (2.01) but with clear distance. Germany shows competitive disadvantage with RTA values of -0.18 in 2017. Ireland also shows competitive disadvantage which is indicated by a RTA value of -4.53 in 2017.

All considered countries except Germany and Ireland show a positive development of RTA values over time. Thereby the positive development of the countries resulted primarily of an increasing relative export advantage then from a decreasing import disadvantage. In the case of the Czech Republic, and France as well the relative export advantage as well as the relative import disadvantage improved over the observed time period. In contrast the negative development of the German relative competitive advantage results of a worsening in relative export advantage and a simultaneous increase in relative import disadvantage.

Table 3. International competitiveness indicators for non-concentrated milk – 2001–2017

	RXA		RMP		RTA		Trend	XS (%)	
	2001–2009	2010–2017	2001–2009	2010–2017	2001–2009	2010–2017	2001–2017	2001–2009	2010–2017
NZ	6.09	13.23	0.28	0.30	5.81	12.93	+	1.25	2.68
BY	3.29	12.55	0.08	0.95	3.20	11.60	+	0.53	2.21
CZ	3.25	4.50	1.14	0.89	2.11	3.61	+	2.62	4.00
PL	1.89	3.17	0.36	1.16	1.54	2.01	+	1.80	3.54
FR	3.27	3.29	2.31	2.18	0.96	1.12	+	12.69	9.46
NL	2.20	2.45	1.88	2.02	0.31	0.44	+	7.27	7.84
AR	0.28	0.13	0.22	0.00	0.05	0.13	+	0.12	0.06
US	0.06	0.10	0.02	0.02	0.04	0.09	+	0.61	0.95
DE	3.56	2.49	2.30	2.68	1.26	-0.18	-	27.12	17.87
IE	1.16	1.30	4.06	5.83	-2.90	-4.53	-	1.21	0.92

Source: Authors' own calculation based on Trademap, 2019.

Regarding the export market share we can see increases in the countries with an increasing trend in RTA values. Exceptions of this are France and Argentina. The strongest decrease in market share can be observed in Germany where the share on worldwide export markets fell from 27.12% in the 2001–2009 period to 17.87% in the 2010–2017 period. Despite this decrease Germany still has the highest market share of all considered countries in this product group in the 2010 to 2017 period.

Concentrated milk and cream

With a volume of € 16.7 bn. the worldwide export market of concentrated milk and cream is the second largest of the considered product groups. As can be seen in Table 4 we find relative competitive advantage in all considered countries in both observed time periods. The highest RTA values again can be seen in New Zealand (205.09) and Belarus (16.03), while the lowest are observed in Germany (0.70) and the Czech Republic (0.47). Within this product group we see increasing relative competitive advantage in all countries except Ireland, Poland and the Czech Republic.

Except for Germany the increasing relative competitive advantage results from an increased relative export advantage, which can also be seen in increasing shares on the international export markets for concentrated milk products except in Germany and France. Against this, the relative export advantage of Germany decreased very slightly from 1.02 in the first period to 0.97 in the 2010–2017 period. Hence, the increasing relative competitive advantage results from an improvement in relative import disadvantage. This is also the case in the France where the relative export advantage decreased while the relative import disadvantage improved.

Table 4. International competitiveness indicators for concentrated milk – 2001–2017

	RXA		RMP		RTA		Trend	XS (%)	
	2001–2009	**2010–2017**	**2001–2009**	**2010–2017**	**2001–2009**	**2010–2017**	**2001–2017**	**2001–2009**	**2010–2017**
NZ	122.44	205.86	0.48	0.77	121.95	205.09	+	18.58	26.88
BY	10.66	16.15	0.16	0.12	10.50	16.03	+	1.67	2.82
AR	8.09	8.32	0.10	0.06	7.98	8.26	+	3.15	3.18
NL	3.13	4.69	2.75	1.06	0.38	3.63	+	10.05	13.86
IE	2.31	2.34	0.46	0.75	1.85	1.59	-	2.34	1.68
FR	1.50	1.77	0.53	0.43	0.97	1.34	+	6.26	5.32
US	0.54	0.93	0.03	0.04	0.51	0.90	+	4.99	8.04
PL	2.99	1.47	0.25	0.81	2.74	0.66	-	2.40	1.64
DE	1.02	0.97	0.40	0.28	0.62	0.70	+	9.68	7.85
CZ	1.36	0.57	0.09	0.10	1.27	0.47	-	0.98	0.52

Source: Authors' own calculation based on Trademap, 2019.

Buttermilk, curdled milk, cream, kefir and yoghurt

The worldwide export market of buttermilk, curdled milk, cream, kefir and yoghurt amounted to € 4 bn. in 2017. Also here New Zealand and Belarus show the highest RTA values in 201 (see Table 5). They are followed by France and Germany with RTA values of 5.06 and 2.15 in the 2010 to 2017 period. Relative competitive disadvantage indicated by negative RTA values can be seen in the Netherlands (-0.94) and Ireland (-3.19).

While New Zealand, Poland and Ireland show decreasing competitive advantage in this product group, the other countries show increasing RTA values over the observed time period. Also here the increase in competitive advantage in Germany results from a stronger decrease in RMP values then in RXA values. In contrast to that the other countries with a positive development in RTA values show an increasing relative export advantage. Though the export market shares of Germany and France are decreasing over the observed time, they are still the highest by far in the 2010 to 2017 period with 20.31 and 15.8%.

Table 5. International competitiveness indicators for buttermilk, curdled milk, cream, kefir and yoghurt – 2001–2017

	RXA		RMP		RTA		Trend	XS (%)	
	2001–2009	2010–2017	2001–2009	2010–2017	2001–2009	2010–2017	2001–2017	2001–2009	2010–2017
NZ	15.37	14.41	0.34	0.25	15.03	14.16	-	3.09	2.92
BY	1.30	10.69	2.86	1.97	-1.56	8.72	+	0.23	1.85
FR	4.83	5.92	1.48	0.85	3.35	5.06	+	17.53	15.80
DE	3.04	2.92	1.03	0.76	2.01	2.15	+	24.18	20.31
PL	2.78	3.05	0.44	1.24	2.34	1.80	-	2.50	3.32
CZ	1.83	2.06	1.88	1.58	-0.05	0.48	+	1.49	1.86
AR	0.28	0.33	0.55	0.02	-0.27	0.31	+	0.12	0.13
US	0.07	0.20	0.04	0.06	0.03	0.14	+	0.73	1.82
NL	0.62	0.85	2.09	1.79	-1.47	-0.94	+	2.18	2.85
IE	1.61	1.91	3.76	5.10	-2.15	-3.19	-	1.70	1.36

Source: Authors' own calculation based Trademap, 2019.

Whey

Whey and whey products had a worldwide export value of € 4 bn. in 2017. With RTA values of 53.91 and 7.4 New Zealand and Belarus do also show the highest relative competitive advantage in this product group (see Table 6). They are followed by Argentina (5.64) and France (3.20). Germany also shows relative competitive advantage with a RTA value of 0.82.

Only the Netherlands show competitive disadvantage in this product group as the RTA value of -1.93 indicates. Thereby the relative competitive advantage decreased in Ireland, Germany, Czech Republic over the observed time. With the exception of the Czech Republic, export market shares in these countries also declined over time. In contrast, the market shares of many other countries, in particular the USA, rose, overtaking Germany in terms of market share as a result of the increase. In the 2010–2017 period, the US market share was 16.06%, while the German exports of whey amounted to 13.36% of the worldwide exports.

Table 6. International competitiveness indicators for whey – 2001–2017

	RXA		RMP		RTA		Trend	XS (%)	
	2001–2009	2010–2017	2001–2009	2010–2017	2001–2009	2010–2017	2001–2017	2001–2009	2010–2017
NZ	46.88	58.85	1.39	4.94	45.49	53.91	+	8.24	10.83
BY	1.05	7.77	0.15	0.47	0.89	7.30	+	0.17	1.41
AR	3.61	5.79	0.38	0.15	3.23	5.64	+	1.47	2.22
FR	4.58	4.63	1.48	1.43	3.10	3.20	+	16.97	12.78
PL	2.97	4.01	0.58	0.80	2.38	3.21	+	2.48	4.39
IE	3.90	4.32	1.17	2.34	2.73	1.89	-	4.03	2.97
US	1.48	2.04	0.63	0.53	0.85	1.51	+	12.94	16.06
DE	1.94	1.77	0.81	0.95	1.14	0.82	-	16.95	13.36
CZ	1.05	1.15	0.33	0.54	0.72	0.61	-	0.79	1.06
NL	3.18	2.86	6.35	4.79	-3.17	-1.93	+	10.19	9.00

Source: Authors' own calculation based on Trademap, 2019.

Butter

In 2016 butter worth € 8.8 bn. was exported worldwide. At 155.75 and 12.64, New Zealand and Belarus also have the highest RTA values for butter, as in the other product groups under review (see Table 7). They are followed by Ireland (13.67) and the Netherlands (2.98). Germany, with an RTA value of -0.41, has relative competitive disadvantages for butter, as do the Czech Republic (-1.21) and France (-1.39). In contrast to Poland and the Czech Republic, however, the development of relative international competitiveness in Germany over time is positive, which can be attributed both to an increase in relative export advantages and to a decrease in import disadvantages. With the exception of the Czech Republic and France, all the countries considered show an increasing relative export advantage while it is decreasing in the Czech Republic and constant in France.

At 23.73%, New Zealand has the highest export market share in the period from 2010 to 2017, followed by the Netherlands (14.37%). Compared to the period from 2001 to 2009, however, this has declined slightly by 0.18 percentage points from 14.55% to 14.37 percentage points. Germany increased its export market share for butter by 1.38 percentage points from 6.66% in the period 2001 to 2009 to 8.04% in the period 2010 to 2017.

Table 7. International competitiveness indicators for butter – 2001–2017

	RXA		RMP		RTA		Trend	XS (%)	
	2001–2009	2010–2017	2001–2009	2010–2017	2001–2009	2010–2017	2001–2017	2001–2009	2010–2017
NZ	104.66	156.09	0.26	0.33	104.40	155.75	+	17.30	23.73
BY	15.77	24.23	0.18	0.10	15.59	24.13	+	2.45	4.18
IE	10.79	14.60	0.61	0.93	10.18	13.67	+	10.31	9.49
NL	4.79	4.83	2.12	1.85	2.66	2.98	+	14.55	14.37
AR	1.34	1.91	0.11	0.08	1.23	1.83	+	0.55	0.78
PL	1.74	1.86	0.37	0.87	1.36	0.99	-	1.49	2.10
US	0.14	0.29	0.14	0.13	-0.01	0.16	+	1.23	2.60
DE	0.68	1.00	1.97	1.41	-1.30	-0.41	+	6.66	8.04
CZ	1.30	0.27	0.91	1.49	0.39	-1.22	-	0.96	0.25
FR	1.49	2.00	2.88	3.39	-1.39	-1.39		6.20	5.96

Source: Authors' own calculation based on Trademap, 2019.

Cheese and curd

With an export volume of € 27.2 bn., cheese and curd were the most important product groups in terms of value. As in the other product groups, New Zealand and Belarus have the highest RTA values with 20.11 and 12.64 respectively (see Table 8). However, New Zealand has declined over time, while Belarus has increased over time. The Netherlands are following in third place with an RTA value of 2.94. With the exception of Germany and the Czech Republic, all the countries considered have relative international competitiveness in cheese and curd. Negative RTA values of -0.47 and -0.78 indicate relative competitive disadvantages for Germany and the Czech Republic.

Table 8. International competitiveness indicators for cheese and curd – 2001–2017

	RXA		RMP		RTA		Trend	XS (%)	
	2001–2009	2010–2017	2001–2009	2010–2017	2001–2009	2010–2017	2001–2017	2001–2009	2010–2017
NZ	22.94	20.82	0.43	0.70	22.51	20.11	-	4.42	4.06
BY	6.03	12.99	0.20	0.35	5.83	12.64	+	0.97	2.24
NL	4.62	4.32	1.18	1.39	3.44	2.94	-	14.11	12.97
FR	4.27	4.54	1.40	1.63	2.83	2.91	+	15.85	12.54
IE	2.90	4.23	1.27	1.92	1.63	2.31	+	2.93	2.95
AR	1.51	2.02	0.18	0.14	1.33	1.88	+	0.61	0.79
PL	1.91	2.25	0.33	0.84	1.58	1.41	-	1.68	2.49
US	0.16	0.47	0.41	0.34	-0.26	0.13	+	1.53	4.22
DE	1.77	2.04	2.55	2.51	-0.79	-0.47	+	15.65	15.10
CZ	0.51	0.72	1.06	1.50	-0.55	-0.78	-	0.39	0.66

Source: Authors' own calculation based on Trademap, 2019.

With the exception of the Netherlands, all countries show a relative export advantage that increased in the second period. The higher relative import disadvantages in Germa-

ny and the Czech Republic ultimately led to negative relative international competitiveness. However, the relative competitive disadvantages of Germany has decreased over time due to an increase in relative export advantage. In addition to New Zealand, the Netherlands, Poland and the Czech Republic also show a relative international competitiveness decline over time. While France still had the highest export market shares for cheese and curd in the period from 2001 to 2009 at 15.85%, it was replaced at the top by Germany in the period from 2010 to 2017. While France's market share fell by 3.31 percentage points to 12.54%, Germany's market share remained at a similarly high level as in the period from 2001 to 2009 due to a minimal minus of 0.55 percentage points at 15.10%.

4.4 Aggregated International Competitiveness of the Dairy Industries

Total global exports across the product groups considered amounted to € 69 billion worldwide in 2017. Similar to the individual product groups, New Zealand and Belarus have the highest RTA values at 89.33 and 14.71 respectively (see Table 9). This is followed by Argentina (2.96) and Ireland (2.13). With an RTA value of 0.17, Germany also has relative competitive advantages across all product groups.

Table 9. International competitiveness indicators for all considered product groups – 2001–2017

	RXA		RMP		RTA		Trend	XS (%)	
	2001–2009	2010–2017	2001–2009	2010–2017	2001–2009	2010–2017	2001–2017	2001–2009	2010–2017
NZ	59.56	90.21	0.45	0.88	59.11	89.33	+	9.17	12.41
BY	7.39	15.14	0.34	0.43	7.06	14.71	+	1.15	2.52
AR	2.76	3.04	0.19	0.09	2.57	2.96	+	1.10	1.18
IE	3.24	4.27	1.45	2.14	1.78	2.13	+	3.28	2.96
FR	3.15	3.45	1.41	1.45	1.74	1.99	+	12.19	9.79
NL	3.62	3.36	2.07	1.63	1.55	1.73	+	11.37	10.36
PL	2.31	2.27	0.34	0.89	1.98	1.38	-	1.98	2.52
US	0.30	0.59	0.22	0.19	0.08	0.41	+	2.85	5.29
DE	1.69	1.71	1.65	1.53	0.05	0.17	+	15.08	12.94
CZ	1.23	1.16	0.80	0.98	0.43	0.18	-	0.95	1.06

Source: Authors' own calculation based on Trademap, 2019.

None of the countries considered has relative competitive disadvantages across all product groups. In addition, with the exception of Poland and the Czech Republic, all the countries surveyed show a positive development in relative international competitiveness over time. At 12.94%, Germany has the highest export market share in the period from 2010 to 2017, ahead of New Zealand (12.41%). However, in contrast to New Zealand's share, Germany's market share has declined over time.

5 Discussion

The analysis of international competitiveness using external trade data has been based on widely used methods in the literature. Weaknesses in older studies were bypassed. Thus, on the one hand, the problem of aggregation was circumvented by the fact that it was not just the dairy industry as a whole that was the subject of the investigation, but also individual product groups. On the other hand, the survey was carried out over a period of 17 years, which shows the development of the international competitiveness of the individual countries for the various product groups over time.

The results show, there is no "international competitiveness" in the world dairy market. The competitive position of the individual states depends much more on the different products. There are two exceptions across all products: New Zealand and Belarus. Both countries show a high degree of international competitiveness across all product groups. For the other states considered, a significantly more differentiated picture emerges. Germany has competitive advantage in buttermilk, cream and yoghurt, whey and concentrated milk products. Thereby the highest RTA-values can be observed at buttermilk, cream and yoghurt, followed by whey and concentrated milk products. The negative RTA values at non-concentrated milk, butter and cheese and curd, however, indicate competitive disadvantage. Over the observed time period we can identify increasing competitive advantage in concentrated milk products, buttermilk, cream and yoghurt, butter, cheese and curd, while non-concentrated milk products and whey show a negative trend in RTA values over the observed time period.

However, the composition of the competitive advantage has to be considered more closely. Except for butter and cheese and curd we can see a decreasing relative export advantage. Over all product groups, the German dairy industry increased its relative export advantage marginally by 0.02 points to 1.71, while the relative import advantage increased by -012 points to 1.53 during the observed time period. The positive trend in competitive advantage in Germany is thus largely due to an increasing relative import advantage. In contrast the competitive advantage and its trend over all product groups in New Zealand, Belarus, Argentina, France, Ireland, the Netherlands and the US results from an increasing relative export advantage, indicating higher international competitiveness on international export markets.

Figure 2. Development of German RTA-values in different product groups

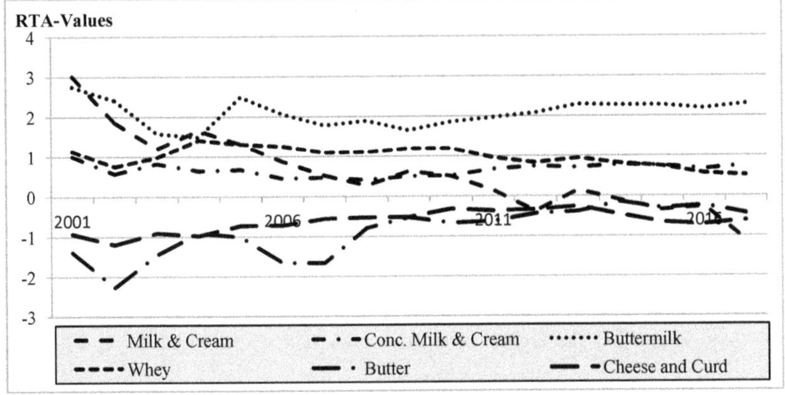

Source: Authors' own depiction and calculation based on Trademap, 2019.

What can we take from these results and how can we explain them? On the one hand, we can see that the main reason for the development of market shares and the relative export advantage in Germany is a relatively stronger increase in competitor exports. Although exports in Germany have increased (see Table 1), exports by competitors and the export market for dairy products have increased more strongly overall, so that the market share across all product groups is declining to the advantage of competitors (see also Figure 3). According to our definition of competitiveness, which also includes the aspect of gaining or at least maintaining market share, Germany's competitiveness can therefore be classified as worse than that of its competitors despite the increase in exports over time.

In addition, the slight decline in the export ratio in the last few years in the German dairy industry shows that, despite the increase in absolute terms, there may well have been difficulties in placing products on the international markets (Janze *et al.*, 2018). How does this fit together? The high market shares in absolute terms indicate that the German dairy industry is highly competitive, but this is primarily the result of the past. Competitiveness has not deteriorated, but competitors have caught up and relative competitiveness has deteriorated. In the future, it is also likely to be even more difficult for the German dairy industry to maintain or expand market shares whereby the reasons

therefore can be seen on the production site. One of the main reasons for this is, for example, the stricter rules on fertilization. The regions with intensive milk production (>2,000 kg milk/ha) were responsible for 78.5% of the growth in milk production in 2015. In the future, it will be much more difficult to expand milk production here, as more land will be needed for the disposal of the manure produced, which will lead to increasing competition with other forms of production for the land and thus to rising land prices (Meyer and Theuvsen, 2017). This will increase production costs of each additional kg, which will have a negative impact on competitiveness.

On the other hand, we see the lowest production costs in the countries in question at 20 to 30 US dollars per 100 kg of milk in New Zealand and Argentina. At 30 to 40 US dollars, Belarus, the USA and Poland follow. They are followed by Ireland, the Czech Republic, Germany and the Netherlands with production costs of 40 to 50 US $ per 100 kg. France has the highest production costs with 50 to 60 US-$ per 100 kg (IFCN, 2013).

Moreover, the projections of the FAO and OECD predict a production growth of milk by 1.8% p.a. whereby 73% of production growth takes place in India and other Asian countries. Coincident the growth of production of fresh milk products is forecasted higher than that of concentrated milk products, butter and cheese (OECD and FAO, 2016). Whereas milk production in Germany is taking place against the background of increasingly stringent requirements with regard to environmental protection and animal welfare, as well as increasing demands by the food retailing industry and consumers, for example in GMO-free feeding. These aspects further reduce the international competitiveness of German dairy products in international markets, except those compensating for these standards.

What are the management implications of the results for the dairy industry in Germany? Against the background of the increasing demands on milk production in Germany, which in most cases will not be compensated for on the international markets, and the forecast development of milk production, the internationalization strategy of dairies should come to the fore in the future. Meyer *et al.* (2019) show in their study that the type of internationalization strategy is an essential success factor for the economic success of German dairies. In their analysis of 18 German dairies, they show that the companies that are significantly more economically successful in opening up foreign markets not only by simply exporting dairy products, but also by accepting and processing milk in other countries. On the other hand, company size, for example, has no significant influence on the economic success of companies. Other studies, such as those by Jürgens *et al.* (2015), also show that the added value of dairies with simple exports is very limited. According to the literature, the pure export strategy can only be successfully implemented with a competitive advantage in the home country (Grant and Nippa, 2006). As mentioned above, this is reduced by the increasing requirements. Against this backdrop, the German dairy industry should, on the one hand, focus on exports of highly refined and special products with high added value, which can compensate for this disadvantage. However, it is also obvious that this strategy is not a solution for all companies due to the product portfolio, let alone for all milk exports. On the other hand, it should examine potential development steps on foreign markets in order to open up the possibility of offering a different product range through market proximity and eventually better framework conditions. Though there are some worldwide operating dairy companies like FrieslandCampina and Nestle, there are just a few private owned German dairies, which are operating factories worldwide, or at least on different continents. For future developments, it will be more important where dairies produce, when exports from the established production regions are of little significance compared to those of other regions when it comes to gain and maintain market share in future milk markets.

On first glance, one aspect might not be fully intuitive, which is the relatively low competitive advantage of Germany in many product groups and over all observed product groups in total, though it has high market shares in the considered product groups. Some studies show similar developments for Germany, for all milk products or for specific product groups (Bojnec & Fertö, 2014). However, this would suggest lower export shares. From a theory's perspective, the high export market shares themselves imply a high degree of competitive advantage when referring to literature. However, in this study, relative competitive advantage was analyzed.

The results clearly indicate New Zealand to be highly competitive (Fahlbusch, 2014). Compared to other countries, e.g. New Zealand, the export of dairy products plays a very underling role in German economy. In 2017 the share of dairy products on total exports in Germany was 0.6%, while the share of the dairy exports on total exports amounted to 23.6% in New Zealand for example (Trademap, 2018). This fact of course results in relatively low relative export advantage of German dairy industry compared to other German industries and other countries considered, where dairy exports play a more important role in overall exports.

Buckley *et al.* (1988) pointed out a general disadvantage in all trade data-related key figures in that they disregard the profitability of the industry or its associated companies. This is a problem when assessing international competitiveness, as market shares may have been bought at the expense of companies' profitability. In this way, high export shares may point to a strong competitive position, even if it is at the expense of the profitability of the companies, which in turn, diminishes international competitiveness. Although the profitability of the industries covered in the individual countries is not part of this work, it cannot be ruled out that such effects are included in the data. However, these are reduced by the long observation period and the division into two different periods, as long as they still play a role.

Frohberg and Hartmann (1997) emphasized in particular, the foreclosure of the domestic market can sometimes cause considerable distortions in the RMP values. If the domestic market is protected by tariffs or import bans, the RMP value is lower, which results in a higher RTA value, which in turn suggests a higher degree of international competitiveness than it actually is. When considering the protection of the domestic market, there are different approaches in the individual countries considered. For example, Argentina, the United States of America and New Zealand use ad valorem tariffs to protect their markets. The European Union, in turn, relies on a value-based taxation of the imported quantity (non-ad valorem tariff) (WTO 2018). In the case of Argentina, for example, the ad valorem tariffs actually applied range between 12.8% (milk) and 23.6% (concentrated milk products) in 2016. New Zealand, on the other hand, levied ad valorem tariffs of 0% in 2016 (milk, butter and cheese and curd) to 3.3% (Product Groups buttermilk, cream and yoghurt and whey).

The United States, on the other hand, imposed both ad valorem tariffs and non-ad valorem tariffs on the various product groups. The ad valorem tariffs in 2016 ranged between 0% (milk) and 18.5% (buttermilk, cream and yoghurt): In addition, there were non-ad valorem tariffs for the various products (WTO, 2018).

Against the backdrop of low protection of the domestic market from foreign dairy products, the already high international competitiveness of the New Zealand dairy industry

is to be assessed as even higher than that of the other countries considered. For Belarus, no information is available

Against the background of the trade barriers mentioned here, the importance of free trade agreements can be recognized both bilaterally and multinationally. Thus, the EU member states are by far the most important trading partners in the European internal market with its open borders for the European countries considered here. Mexico is also the most important trading partner of the United States, which is linked by the North Atlantic Free Trade Agreement (NAFTA). New Zealand has a Free Trade Agreement with China, which is also the main purchaser of New Zealand dairy products. Belarus, Russia and Kazakhstan are members of the Eurasian Economic Union. 93.7% of Belarus' total exports go to both countries. These results are supported by Couillard and Turkina (2015), who have studied the positive effects on the production of competitive states. Since we have only examined net exporters of dairy products in this study, we assume that they are competitive with other countries and that these results are therefore also valid here. On the other hand, the exports from European countries to Russia, which in the course of the trade dispute in 2012 have partly fallen to zero, show what distortions in these relations can have for repercussions.

Irrespective of the trade agreements between the countries analyzed, we can see clear trends in market shares on the global export market for dairy products. Figure 3 shows the development of the market shares of the countries in question in index form. The market share in 2001 corresponds to an index value of 100. It is clear that Belarus has been able to increase its market shares most significantly over time, although in absolute terms they are at a low level (cf. Table 1 in the Appendix). The USA and New Zealand also increased their market shares over time, although New Zealand's market share is significantly higher. Argentina's market share fluctuates over time and declines significantly from 2013 onwards. In the European countries, the two Eastern European countries Poland and the Czech Republic were able to increase their export market shares over time, whereas the export market shares of the Western European countries Germany, France, Ireland and the Netherlands (red lines) declined over time but at a significantly higher level.

Figure 3. Development of export market share indexes and the value of world dairy exports

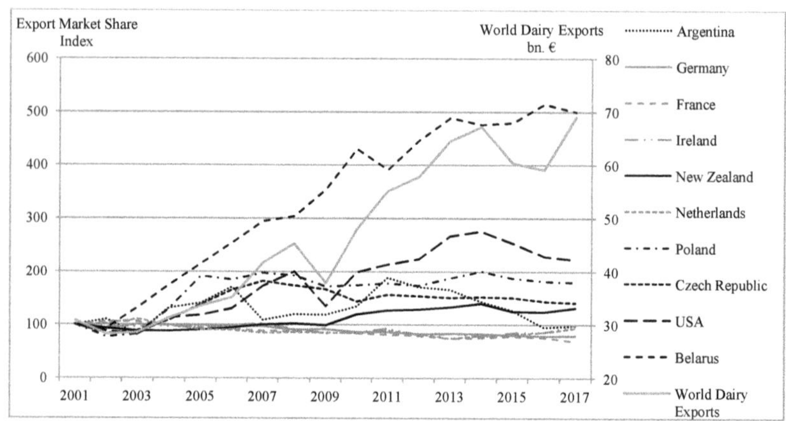

Source: Authors' own depiction and calculation Trademap, 2019.

The result is a little over-hasty, as one would have expected that the gradual increase in the quota from 2009/2010 until its abolition in 2015 would also increase the market shares of the European states. The result does not mean, however, that exports in these countries have not risen; rather, the entire export market has grown faster than exports from Western European countries, which has led to falling index values over time.

The results show a sharp increase in international competitiveness for milk products from Belarus. These are worth to have a closer look on. According to the considered trade data these results are not surprising as Belarus shows a strong increase in trade balance, as well an in the share of dairy products on total exports (2% in 2001, 6.3% in 2016). But what are the drivers behind that and can we trust in these results? According to the FAO Belarus increased its milk production by 51.9% to 7.31 m. tons in (FAOSTAT, 2019). The results of Ramanovich (2010) show lower production costs in milk production in Belarus compared to western and eastern European countries which are implicating a competitive advantage of Belarus. And also his analysis of foreign trade data is underlining this result. Nevertheless we have to be aware that the foreign trade of Belarus is nearly completely based on the trade with Russia (cf. Table 2). This might be an implication for lower quality standards as the ratio is traditionally high. Otherwise one could argue if this result is based on the trade war between the European Union and Russia. Nevertheless, assuming that we trust the many sources and their data, we can say that Belarus's competitiveness has indeed increased over time.

Given the import ban by Russia, the high values might reflect the extraordinary situation, that Belarus has developed into an unofficial trading platform between the EU and Russia, substituting the banned direct imports from the EU to Russia, where EU agri-food products are unofficially delivered through Belarus. However, there is no official information available.

Even though we have tried to overcome some limitations of similar, earlier studies within this investigation, there is also a need for further research in this area. On the basis of foreign trade data, we were able to quantify the international competitiveness of the individual countries and their dairy industry. However, the detailed reasons for this result are not revealed. Melitz (2003), for example, points out that productivity is a factor for export success and thus for increasing market shares. Whether the reason for the competitiveness of the individual countries can be traced back to higher productivity can only be assumed because there is no data basis for this. However, productivity is likely to determine the degree of competitiveness vis-à-vis the other country, at least for countries with comparable framework conditions. In addition, other factors, such as the natural environment, can influence competitiveness. This aspect is likely to apply primarily to New Zealand. At this point, future studies should follow, which examine the causes for the competitive advantages of the individual states and their dairy industry.

6 Conclusion

The work shows significant differences in the international competitiveness of individual countries in the various product groups. Across all product groups, New Zealand and Belarus are the most competitive. Based on foreign trade data, the competitive advantage of the German dairy industry has slightly increased over the past 17 years over all product groups, whereby there are great differences between the considered product groups. While Germany shows competitive advantage in concentrated and non-concentrated milk products, buttermilk, cream and yoghurt and whey, there is competitive disadvantage in milk, butter and cheese and curd. However, the positive trend in RTA values over the observed time mainly results from an improved relative import advantage and not from an increasing relative export advantage. This indicates that it increasingly difficult for the German dairy industry to compete with other exporters on international markets. This result is also supported by the market share in the global export market of the product groups, which, in addition to the exceptions in butter, has in some cases fallen considerably. Nevertheless, the market shares in the individual product groups, which are in some cases high compared to other competitors, speak for a strong competitive position in the past. However, the result is not due to a worsening of the situation in Germany itself, but important competitors have caught up and competition in international markets for dairy and milk products has intensified, which reduces the relative international competitiveness of the German dairy industry. This result is also supported in view of the other Western European countries which, like Germany, have high absolute market shares, but have not been able to increase or maintain these over time in contrast to other competitors, despite an increase in the milk quota as part of its abolition. Furthermore, the relatively low values of competitive advantage also refer to other strong export oriented and competitive industries in Germany, especially automobiles and machine engineering.

Nevertheless, the results of the study, in addition to the forecast of the worldwide growth in milk production and increasing requirements in German milk production, imply that it will be increasingly difficult for Germany dairies to place their products on the world market in the future, especially in bulk markets. Against this background, it will become increasingly important for dairy companies to produce at the right places in the future in order to successfully gain market share in the growing global milk market. Some German dairies are already represented worldwide with production sites. Large dairy companies such as FrieslandCampina and Lactalis have been successfully implementing this strategy for much longer (Meyer *et al*,. 2019). Especially for the cooperative dairies, this strategy is likely to be very challenging due to the ownership structure. In the future, however, apart from niche products, this strategy should be much more successful than simply exporting dairy products.

With a view to the future, the image of competitors will also change. While Germany currently competes primarily with the other European countries for the European market, competition with New Zealand and the United States in particular for new and existing markets, especially China and Asia, will increase in the future.

References

AMI (2015). AMI-Marketbalance Milk 2015. Agrarmarkt Information mbH (AMI).

AMI (2018). AMI-Marketbalance Milk 2018. Agrarmarkt Information mbH (AMI).

Baade, D. (2007). Demografischer Wandel und international Wettbewerbsfähigkeit Deutschlands. Dissertation, University of Hannover, Hannover.

Balassa, B. (1962). *Recent Developments in the Competitiveness of American Industry and Prospects for the Future*. Yale University Press.

Bellora, C., Emlinger, C., Fouré, J., Guimbard, H. (2017). Research for AGRI Commitee, EU – UK agricultural trade: state of play and possible impacts of Brexit. European Parliament, Policy Department for Structural and Cohesion Policies.

Bojnec, S. & I. Fertö (2014). Export competitiveness of dairy products on global markets: the case of the European Union countries. *Journal of Dairy Science* 97(10): 6151–6163.

Brümmer, B. (2019). Auswirkungen des Brexit auf den Handel mit Milch- und Milchprodukten. Public lecture series of the Faculty of Agriculture – Milchtrends, University of Göttingen.

Buckley, P.J., Pass, C.L., Prescott, K. (1988). Measures of international competitiveness: A critical survey. *Journal of Marketing Management* 4(2): 175–200.

Couillard, C., Turkina, E. (2015). Trade Liberalization: The Effects of Free Trade Agreements on the Competitiveness of the Dairy Sector. *The World Economy*, 38(6): 1015–1033.

Destatis (2008–2016). Concentration in the manufacturing industries 2008–2016. Statistical report of the federal statistical office (Destatis).

Destatis (2019a). Employees and turnover in the German manufacturing industry from 2008 to 2016. Genesis-Online database, Federal Office for Statistics, available at: https://www-genesis.destatis.de/genesis/online (accessed 20 April 2019).

Destatis (2019b). Exports of German dairy and dairy products from 2008 to 2016. Genesis-Online database, Federal Office for Statistics, available at: https://www-genesis.destatis.de/genesis/online (accessed 10 December 2018).

Dialer, D., Lichtenberger, E., Neisser, H. (2010). *Das Europäische Parlament – Institutionen, Visionen und Wirklichkeit.* Innsbruck University Press, 1st edition.

Drescher, K. and Maurer, O. (1999). Competitiveness in the European Dairy Industries. *Agribusiness* 15(2): 163–177.

EU Commission (2016). EU agricultural outlook: Prospects for the EU agricultural markets and income 2016–2026 (12/2016). EU Commission, DG Agriculture and Rural Development, available at: https://ec.europa.eu/agriculture/sites/agriculture/files/markets-and-prices/medium-term-outlook/2016/2016-fullrep_en.pdf. (accessed 15 April 2019).

EU Commission (2019). Russian import ban on agricultural products. EU Commission, DG Agriculture and Rural Development, available at: https://ec.europa.eu/agriculture/russian-import-ban_fr. (accessed 15 April 2019).

Eurostat (2019). Cows' milk collection and production obtained – monthly data. Eurostat database, available at: https://ec.europa.eu/eurostat/de/data/database (accessed 20 February 2019).

Fahlbusch, M. (2014). Price Formation and the Measurement of Market Power on the International Dairy Markets. PhD thesis at the chair of Agricultural Market Analysis, University of Göttingen.

FAOSTAT (2019). Production of milk, whole fresh cow in Belarus from 2001 to 2017. FAOSTAT, Food and Agriculture Organization of the United Nations, available at: http://www.fao.org/faostat/en/#data (accessed 10 April 2019).

Fedoseeva, S. (2016). Russian agricultural import ban: Quantifying losses of German agrifood exporters. 56th annual conference of the German Association of Agricultural Economists, 28–30 September, Bonn, Germany.

Fertö, I. and Hubbard, L.J. (2003). Revealed Comparative Advantage and Competitiveness in Hungarian Agri-Food-Sectors, *The World Economy* 26(2): 247–259.

Feurer, R. and Chaharbaghi, K. (1994). Defining Competitiveness: A Holistic Approach. *Management Decision* 32: 49–58.

Frohberg, K. and Hartmann, M. (1997). Comparing Measures of Competitiveness. Discussion paper No. 2, Institute of Agricultural Development in Central and Eastern Europe, Leibnitz Institute of Agricultural Development in Transition Economics.

Guillouzo, R. and Ruffio, P. (2005). Internationalization of European dairy co-operatives. *International Journal of Co-operative Management* 2(2): 25–35

Grant, R.M., Nippa, M. (2006). *Strategisches Management.* Pearson Studium, Hallbergmoos.

Gries, T. and Hentschel, C. (1994). Internationale Wettbewerbsfähigkeit – was ist das? *Wirtschaftsdienst* 74: 416–422.

Hörl, M., Hess, S. (2017). The Export Competitiveness of the European Dairy Industry. Contribution at the XV Congress of the European Association of Agricultural Economists, 28 August – 1 September, Parma, Italy.

Heyder, M., Makus, C., Theuvsen, L. (2011). Internationalization and Firm Performance in Agribusiness: Empirical Evidence from European Cooperatives. *International Journal on Food System Dynamics* 2(1): 77–93.

IFCN (2013). Dairy Report 2013. International Farm Comparison Network.

Ishchukova, N. and Smutka, L. (2013). Revealed Comparative Advantage of Russian Agricultural Exports, *Acta Universitatis Agriculturae et Silviculturae Mendelianae Brunensis* LXI (4): 941–952.

Janze, C., Theuvsen, L., Schmidt, C., Meyer, J., Winkel, C. (2018). Konjunkturbarometer Agribusiness in Deutschland 2018, Ernst & Young, Stuttgart.

Jürgens, K., Fink-Kessler, A., Poppinga, O., Wohlgemuth, M. (2015). Wertschöpfung von Molkereien. MEG Milch Board, Göttingen.

Kirner, L. (2009). Auswirkungen der vollständigen Implementierung des Health-Check auf landwirtschaftliche Betriebe. *Ländlicher Raum,* online-journal of the Austrian Ministry for Agriculture, Forestry, Environment and Water Conservancy, Wien, Austria.

Man, T.W.Y., Lau, T., Chan, K.F. (2002). The competitiveness of small and medium enterprises – A conceptualization with focus on entrepreneurial competencies. *Journal of Business Venturing* 17(2): 123–142.

Martin, L., Westgreen, R., van Duren, E. (1991). Agribusiness Competitiveness across National Boundaries. *American Journal of Agricultural Economics* 73(5): 1456–1464.

Marvel, H.P. and Ray, E.J. (1987). Intraindustry Trade: Sources and Effects on Protection. *Journal of Political Economy* 95(6): 1278–1291.

Melitz, M. J., (2003). The Impact of Trade on Intra-Industry Reallocations and Aggregate Industry Productivity. *Econometrica* 71(6): 1695–1725.

Meyer, J., Theuvsen, L. (2017). Intensive Dairy Farming in Northern Germany: Development and Impact of the New Fertilizer Act. In: Proceedings of Agrarian Perspectives XXVI, Competitiveness of European Agriculture and Food Sectors, Proceedings 26th International Scientific Conference, 13–15 September, Prague: 219–225.

Meyer, J., Feil, J.H., Schaper, C. (2019). Internationalization Strategies in the German Dairy Industry and their Influence on the Economic Performance of Firms. *International Journal of Food System Dynamics* 10(4): 332–346.

Nelson, R. (1992). Recent Wrtings on Competitiveness: Boxing the Compass. *California Management Review* 34(2): 127–137.

MIV (2019). Wohin die Milch fließt. Dairy Industry Association of Germany, available at: https://milchindustrie.de/wp-content/uploads/2018/11/Wohin-die-Milch-flie%C3%9Ft-2018.pdf (accessed 10 March 2019).

OECD, FAO 2016. OECD-FAO Agricultural Outlook (2016–2025). OECD Publishing, Paris.

Poppe, K.J., Wijnards, J.H.M., van der Meulen, B.M.J., Bremmers, H.J. (2007). Struggle for Leadership: The Competitiveness of the EU and US Food Industry, Paper for poster presentation, American Agricultural Economics Association Annual meeting, 29 July – 1 August, Portland, OR, USA.

Rabobank 2018. Global Dairy Top 20 (2018). Raobbank, available at: https://research.rabobank.com/far/en/sectors/dairy/Dairy-top-20-2018.html. (accessed 10 December 2018).

Ramanovich, M. (2010). Zur Bestimmung der Wettbewerbsfähigkeit des weißrussischen Milchsektors: Aussagefähigkeit von Wettbewerbsindikatoren und Entwicklung eines kohärenten Messungsinstruments. Dissertation, Martin Luther University, Halle-Wittenberg.

Serin, V. and Civan, A. (2008). Revealed Comparative Advantage and Competitiveness: A Case Study for Turkey towards the EU, *Journal of Economic and Social Research* 10(2): 25–41.

Scott, L. and Vollrath, T. (1992). Global Competitive Advantages and Overall Bilateral Complementarity in Agriculture – A Statistical Review. *Economic Research Service*, U.S. Department of Agriculture, Washington D.C., USA.

Simo, D., L. Mura & J. Buleca (2016). Assessment of milk production competitiveness of the Slovak Republic within the EU-27 countries. *Agricultural Economics – Czech* 62(10): 482–492.

Theuvsen, L., Janze, C., Heyder, M. (2010). Agribusiness in Germany 2010. On the Way to New Markets. Ernst & Young, Stuttgart.

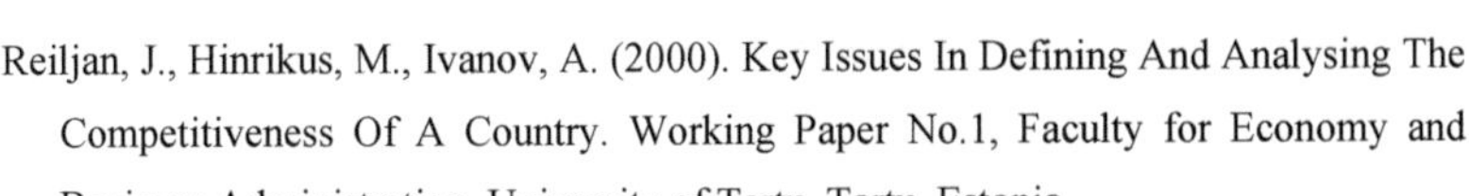

Reiljan, J., Hinrikus, M., Ivanov, A. (2000). Key Issues In Defining And Analysing The Competitiveness Of A Country. Working Paper No.1, Faculty for Economy and Business Administration, University of Tartu, Tartu, Estonia.

Trabold, H. (1995). Die internationale Wettbewerbsfähigkeit einer Volkswirtschaft. *Vierteljahreshefte zur Wirtschaftsforschung* 64(2):169–185.

Trademap (2019). Data of trade for product groups 0401-0406. Inetrnational Trade Center, available at: https://www.trademap.org/Index.aspx (accessed 20 June 2019).

Trail, W.B. and Pitts, E. (1998). *Competitiveness in the Food Industry*. Blackie Academic & Professional.

Utzig, S. (1987). Internationale Wettbewerbsfähigkeit, Unternehmensorganisation und Ordnungspolitik. *Wirtschaftsdienst* 67(8): 417–422.

Vitaliano, P. (2016). Global Dairy Trade: Where Are We, How Did We Get Here and Where Are We Going? *International Food and Agribusiness Management Review* 19(B): 27–36.

Vollrath, T.L. (1987). Revealed Competitiveness for Wheat. Economic Research Service Staff Report No. AGES861030, U.S. Dept. of Agriculture, Economic Research Service.

Vollrath, T.L. (1989). Competitiveness and Protection in World Agriculture. Agricultural Information Bulletin, No. 567, U.S. Dept. of Agriculture, Economic Research Service.

Vollrath, T.L. (1991). A Theoretical Evaluation of Alternative Trade Intensity Measures of Revealed Comparative Advantage. *Weltwirtschaftliches Archiv* 127(2): 265–280.

Weindlmaier, H. (2000). Die Wettbewerbsfähigkeit der deutschen Ernährungsindustrie: Methodische Ansatzpunkte zur Messung und empirische Ergebnisse. *Schriften der Gesellschaft für Wirtschafts- und Sozialwissenschaften des Landbaues e.V.* 36: 239–248.

Winter, M. and Würmann, C. (2012). Wettbewerb an Hochschulen. *Die Hochschule* 2: 6–16.

WTO – World Trade Organization (2018). MFN Applied Tariff, data on applied tariffs for different dairy products. Tariff download facility, World Trade Organization, available at: http://tariffdata.wto.org/ReportersAndProducts.aspx. (accessed 10 December 2019).

Yu, R., Junning, C. and PingSun L. (2008). *The normalized revealed comparative advantage index*. Springer Verlag.

ZMB (2016). ZMB Jahrbuch Milch. ZMB Zentrale Milchmarkt Berichterstattung GmbH.

Appendix

Table 1. Export market shares of considered countries

	AR	DE	FR	IE	NZ	NL	PL	CZ	US	BY
2001	1.0	15.7	12.7	3.4	9.7	12.1	1.3	0.7	2.3	0.5
2002	1.1	14.4	13.4	3.3	9.0	12.3	1.0	0.6	2.0	0.5
2003	0.8	15.7	13.4	3.3	8.5	13.3	1.1	0.6	1.8	0.7
2004	1.3	15.5	12.4	3.3	8.5	12.0	1.7	0.8	2.5	0.9
2005	1.4	15.5	11.9	3.2	8.7	11.0	2.6	1.0	2.7	1.2
2006	1.7	15.2	11.7	3.4	9.1	10.9	2.5	1.2	2.9	1.4
2007	1.1	15.2	11.2	3.5	9.7	10.2	2.7	1.3	3.9	1.6
2008	1.2	14.2	11.1	3.2	9.9	10.4	2.6	1.2	4.5	1.6
2009	1.2	14.3	11.8	2.9	9.6	10.3	2.3	1.2	3.0	1.9
2010	1.4	13.6	10.6	2.9	11.5	10.2	2.3	1.0	4.5	2.3
2011	1.9	13.5	10.3	3.2	12.3	10.8	2.4	1.1	4.8	2.1
2012	1.7	13.0	10.1	2.8	12.4	10.0	2.3	1.1	5.0	2.4
2013	1.7	13.1	9.4	2.9	12.8	9.0	2.5	1.1	6.0	2.6
2014	1.5	13.0	9.6	2.7	13.5	9.5	2.7	1.1	6.2	2.6
2015	1.3	12.6	9.9	2.9	12.1	9.9	2.5	1.1	5.7	2.6
2016	1.0	12.4	9.6	2.9	12.0	10.3	2.4	1.0	5.1	2.8
2017	1.0	12.4	8.7	3.4	12.7	11.3	2.4	1.0	5.0	2.7

Source: Authors' own calculation based on Trademap 2019.

V Internationalization Strategies in the German Dairy Industry and their Influence on the Economic Performance of Firms

Authors: Johannes Meyer, Jan-Henning Feil, Christian Schaper

Published in: Journal of Food System Dynamics Vol. 10 (2019)

Abstract

Growing milk production, stagnating domestic consumption and ongoing liberalization of the worldwide milk market have led to increasing exports of milk and milk products out of Germany. This situation heightens competition amongst German dairies for market share on foreign markets. The German dairy industry, which comprises of some international corporations and many medium sized companies, including both cooperatives and privately owned companies, therefore has to find strategies with which to compete successfully on international markets. This study analyzes the German dairy industry comparing different internationalization strategies and their influence on the firms' economic success. 18 German dairy companies have been analyzed. We identified different internationalization strategies with reference to Perlmutter's EPRG model. To measure economic success, we analyzed annual reports from the dairy companies observed over the years 2010 to 2017 and so calculated different key figures. The influence of different internationalization strategies on economic success is analyzed by a Hausman Taylor estimation where the EBIT-margin is the dependent variable in our model, representing economic success. We found that German dairy industry companies do pursue different internationalization strategies and that these have different influences on the companies' economic success.

1 Introduction

The dairy industry has been facing a process of rapid internationalization due to trade liberalization and regional imbalances on the world milk market (Guillouzo and Ruffio, 2005; Heyder *et al.*, 2011; Vitaliano, 2016). As a consequence, international trade flows between net exporters, such as New Zealand, the United States and the European Union, and net importers, such as China, the Middle East and Africa, have been increasing. At the organizational level, this development has resulted in growing export activity on the part of dairy companies located in net export regions like the United States and the European Union (Vitaliano, 2016). This situation can be illustrated by a look at the German dairy sector, which has faced constant increases in milk production over the course of the last decade. During the same time period, domestic demand has remained more or less at a stalemate, leading to a growing milk surplus. This surplus had to be exported, and German dairies were increasingly forced to look for market opportunities abroad (Heyder *et al.*, 2011).

The relevance of exports is even greater when one considers the physical amounts being exported. In 2016, German dairy companies processed 33.8 million tons of milk, including 2.5 million tons of raw milk imports from neighboring countries. Of this, 16.6 million tons of milk equivalent were exported as cheese, whole or skimmed milk powder and other products, which corresponds to 49.1 % of processed milk (MIV, 2018). Internationalization has therefore become the foremost driver of industry development (Theuvsen *et al.*, 2010). Companies employ different internationalization strategies with which to compete for market share in foreign markets. These internationalization strategies differ from simple export from the country of origin to foreign markets to multinational companies with manufacturing plants and offices all over the world. Dutch and Scandinavian dairies lead in internationalization. Due to their limited domestic market size and high milk production volumes, they were forced to look for marketing opportunities abroad much earlier than German dairies (Theuvsen and Ebneth, 2005; Heyder *et al.*, 2011).

Despite numerous studies in the literature which have investigated internationalization in the dairy sector, there is a research gap; what was the influence of the different internationalization strategies employed by firms on their economic performance? (Guillouzo and Ruffio, 2005; Heyder *et al.*, 2011; Theuvsen and Ebneth, 2005). This

study will fill this gap by analyzing the influence of different internationalization strategies on the economic performance of firms in the German dairy sector.

To analyze the effects of different internationalization strategies on financial performance, we examined 15 leading German dairies as well as two foreign dairies which are operating in Germany. Our sample included cooperatives as well as privately operated dairies. The data used was derived from the annual reports and annual financial statements of the companies under analysis for the years 2010 to 2016. To analyze the influence of different internationalization strategies on the economic performance of firms the different internationalization strategies of German dairies are analyzed, referring to the EPRG model. To measure the influence of the different internationalization strategies on the economic performance of the considered firms, we used a random effects model to analyze our panel data.

2 Theoretical Background

In the literature, internationalization is defined as the transnational transactions of an organization (Fayerweather, 1989). The internationalization of the German dairy sector has been analyzed in previous studies. Theuvsen and Ebneth (2005) analyzed the degree of internalization in cooperatives in the German dairy and meat sector using different uni- and multidimensional key figures. Heyder *et al.* (2011) analyzed the effects of internationalization on economic success in European dairy and meat cooperatives using financial report based key figures while defining internationalization by the Degree of Internationalization (DOI). To measure economic success, they used the variables return on assets (ROA) and return on sales (ROS). Internationalization was measured using the DOI as a multidimensional key figure calculated on foreign sales to total sales and the network spread index (NSI). They found that the degree of internationalization has a significant positive influence on the firm's economic success. Widely used key figures for measuring economic performance are Return on Assets (ROA), Return on Equity (ROE) and EBIT margin (Qian, 2002; Vermeulen and Barkema, 2002; Thomas and Eden, 2004; Heyder *et al.*, 2011; Chaddad and Mondelli, 2013).

However, these studies did not define or distinguish between different internationalization strategies. According to Johnson *et al.* (2016), an internationalization strategy can be defined as the long-term alignment of a firm competing on foreign markets with respect to resources and market shares. The competitive advantage of internationalization results from two opposing effects: advantages resulting from local adaptation and differentiation, and advantages resulting from global standardization (Johnson *et al.*, 2016). With regard to these two dimensions, one can differentiate between four different internationalization strategies in the EPRG model, which was introduced by Perlmutter (1969). He differentiated between ethnocentric, polycentric and geocentric concepts. Later, this model was extended through the introduction of a regiocentric concept (see Figure 1) (Wind *et al.*, 1973).

Figure 1. EPRG model

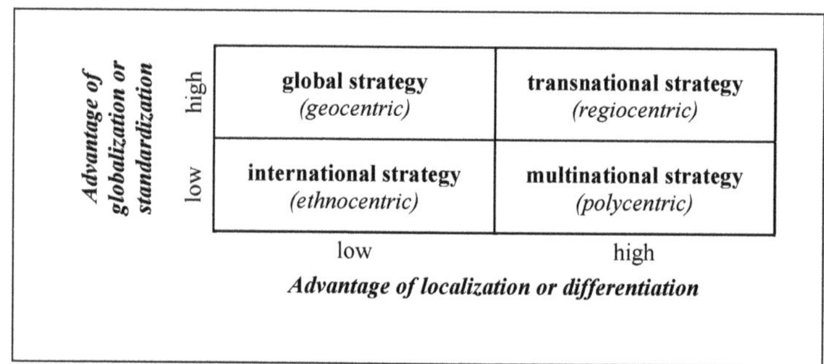

Source: Authors' own depiction based on Holtbrügge and Welge, 2015.

The international strategy (ethnocentric) is also described as an export strategy (Johnson *et al.*, 2016). Subsidiaries, if there are any, are guided by the parent company and seen as additions to international business or as generators of short term profits (Magaziner and Reich, 1985). This strategy can be successfully implemented if there is a competitive advantage in the home country that cannot be achieved in the target countries for the exports (Grant and Nippa, 2006). With multinational strategy (polycentric), subsidiaries can be led by foreign executives and are less strictly coordinated by the parent company. Thus, national strategies can be implemented and greater efficiency achieved through better adaptation to local demand and preferences. However, economies of scope and synergies resulting from internationalization can be restricted when implementing this strategy (Scholl, 1989).

Global strategy (geocentric) focuses on economies of scope. This strategy is also known as global rationalization. Firms try to formalize and standardize structures, processes and resources, while decision making competences are centralized in the parent company. Technology is also transferred from the parent company to its subsidiaries. The advantages that result from realizing economies of scope in this strategy counter the disadvantages of a lack in adaptation to local demand and preferences (Negandhi and Welge, 1984). None of the strategies described so far can combine the advantages of local adaptation and differentiation, on the one hand, with standardization and economies of scale, on the other. The transnational strategy (regiocentric) combines these two advantages of internationalization. Advantages from differentiation and standardization are analyzed for each business activity. The company's global alignment and the simultaneous country- or region-specific treatment of markets combines the advantages of the multinational and global strategies (Scholl, 1989). However, in reality, there is not always such a sharp distinction between the different approaches. As a result, firms often develop regional strategies and, in doing so, combine the global and multinational strategies (Johnson *et al.*, 2016). In our study we will refer to Perlmutter's EPRG model (1969) to define and distinguish the different internationalization strategies employed by the firms considered.

3 Data and Methods

This study is restricted to firm level data. All data used derive from the firms' annual reports. These were either collected from the companies' websites or in the German Bundesanzeiger[1]. As far as possible, annual reports from 2010 to 2017 were collected for 18 dairies. 16 of the dairies are headquartered in Germany, while two dairies are not originally from Germany. However, these two companies process considerable quantities of milk in Germany and have, over time, emerged as major players in the German milk market. As a result, they were also included in this study.

To analyze the influence of different internationalization strategies on firm' economic performance, we first defined the internationalization strategy for each of the firms considered, referring to Perlmutter's EPRG model (1969). Different relative and absolute, unidimensional and multidimensional key figures were used to analyze the firms' financial performance and development of their international businesses in relation to their varying internationalization strategies.

[1] The German Bundesanzeiger publishes the annual financial reports of German corporations (www.bundesanzeiger.de)

To measure the influence of the different strategies on the firms' economic success, a profitability measure has to be defined. A widely used profitability measure is profit margin or return on sales (Qian, 1994; Capar and Katobe, 2003; Li, 2007; Heyder *et al.*, 2011). However, we use Earnings Before Interest and Taxes (EBIT) as the most appropriate profitability measure because two dairies are headquartered in foreign countries and therefore do not refer to the German accounting standards in their annual reports. Different taxes and levels of taxes might therefore bias the results, if we were to take them into account. EBIT is a widely used performance indicator because it separates the firm management's financing decisions from its fundamental earning potential (Becker-Blease *et al.*, 2010; Heyder *et al.*, 2011). To make the data of different sized companies comparable we use the EBIT margin (EM) as the variable to measure the firms' economic performance. It is calculated as the ratio of EBIT and turnover.

$$EBIT\ margin = \frac{\text{EBIT}}{\text{Turnover}} \tag{1}$$

To measure the influence of the different internationalization strategies on economic performance we implemented the dummy variable "Internationalization Strategy". We coded this with "0" if the company pursues the "international strategy" and with "1" if the company pursues the multinational strategy. To isolate the relationship and influence of the different internationalization strategies on the firm's economic performance, it is important to control for other variables that are likely to affect economic performance (Qian, 2002). To isolate the influence of the different internationalization strategies on firms' performance we added proxy variables for the firms' size, debt level, product price, number of brands, as well as dummy variables for the organizational form and for the products the firms are producing.

Our first control variable is size. We therefore use the overall turnover of the firm considered. We use firm size as an independent variable in the model as a surrogate for competitiveness and the firm's advantage within the industry (Qian, 2002; Heyder *et al.*, 2011). As a second control variable we use the debt level, specifically the "Gearing Ratio" which is calculated as the ratio of debt capital to the firm's equity (Hurdle, 1974). For comparison we use liabilities to credit institutes as long term debts. It is a proxy for the ability of a firm to meet long term interest and principal payments on debt. However, it is also a measure for the firm's financial risk, which increases with a higher gearing ratio (Qian, 2002). As a third control variable we implemented the FAO Food Price Index for dairy products, as we expect that profitability will increase ceteris paribus

with increasing product prices. To control for the influence of the firms' organizations we implement the dummy variable "Legal Form" in the model. It differs between "private" (code = 1) and "cooperative" (code = 0) to control for the influence of organizational form on economic performance. This is important, especially in the German dairy industry, which is characterized on the one hand by a large dairy cooperative and, on the other hand, by many medium sized cooperative and private dairies (Theuvsen and Ebneth, 2005). Thus dairies are often linked with poorer economic performance, compared to privately owned competitors. This is due to conflicting goals between the dairy and its members, lack of management competences and so on (Anderson and Henehan, 2005). In order to check the influence of the companies' product portfolio on their profitability, dummy variables for the product categories "fresh milk", "fresh milk products", "dried milk products", "cheese" and "butter" were included in the model.

Lack of any, or in large parts very incomplete data on the quantity of milk consumed and processed as well as other intermediate products, plus there being no complete and comparable data on output quantities, made it impossible to investigate the effects of different productivity ratios on companies' economic success, even if there are clear indications in the literature of their influence (Helpman *et al.*, 2003).

To analyze our unbalanced micro panel data we use the Stata 15 software. As we have endogeneity between the dependent variable and the variable "internationalization strategy", we use the Hausman-Taylor estimator to analyze our data. Due to heteroscedasticity and autocorrelation in the data, we also have clustered standard errors that are robust against them.

$$Y_{it} = \mathrm{X1}_{it}\beta_1 + \mathrm{X2}_{it}\beta_2 + \mathrm{Z1}_i\gamma_1 + \mathrm{Z2}_i\gamma_2 + \alpha_i + \varepsilon_{it} \qquad (2)$$

Thereby Y_{it} it is the dependent variable of Firm i at Time t. $\mathrm{X1}_{it}$ are time-varying and $\mathrm{Z1}_i$ are time-invariant variables that are assumed to be exogenous and not correlated with α_i and ε_{it}. $\mathrm{X2}_{it}$ are endogenous time-variant and $\mathrm{Z2}_i$ are endogenous time-invariant variables as they correlate with α_i, but not with ε_{it} (BALTAGI *et al.*, 2002). Thereby α_i denotes the time-invariant term of the error term, while ε_{it} denotes the remainder component of the error term which is uncorrelated over time (Verbeek, 2004).

4 Results

4.1 Internationalization Strategies in the German dairy Industry

Referring to Perlmutter's EPRG model (1969), we can identify three different internationalization strategies followed by the firms in this study. Ten of the firms—the majority—employ the International Strategy (see Table 1).

Seven are cooperatives, and the other three are privately owned. All ten firms are characterized by the fact that they have processing facilities only in Germany. Some of these firms do have subsidiaries in foreign countries. The number of countries in which they operate subsidiaries ranges from 1 (frischli, Omira, Uelzana, Rücker Wismar) to 10 (DMK) in 2017. These subsidiaries are only engaged in marketing and distribution but do not process milk. On average the firms in this group had subsidiaries in 3 countries including Germany.

Therefore, foreign sales only result from export activities out of Germany. On average the dairies employing the International Strategy had 4.9 brands in 2016, but the number varies from 1 to 11. The foreign sales index (FSI), measured as foreign sales to total sales, was 33.3 % on average in 2017 in this group; it ranges from 0 % at Rücker Wismar to 47.7 % at Rücker Aurich. Two of the firms in the study, one cooperative and one privately owned company, have implemented a global strategy. These firms operate processing facilities in neighboring countries outside Germany. On average these firms have subsidiaries in 7.5 countries. Although some national brands are used in foreign countries, the relevant firms have introduced special brands for their foreign markets. On average these firms have 15.5 national and international brands to adapt to market demands. The short distance from Germany to the neighboring countries ensures comprehensive control for the parent company where decision making competences are retained. In 2017 the average FSI in this group was 51.1 %, whereby the FSI of Zott (56.5 %) was higher than that of Hochwald (45.6 %).

The third group, consisting of five firms, uses the Multinational Strategy. These firms have processing locations in several countries, some of which are far from their domestic market. On average the firms employing the Multinational Strategy had subsidiaries in 22 countries in 2017. These firms manage several brands for the markets they supply. On average this group of firms had 19.8 brands in 2017. The actual number of brands varies from Meggle's 1 to FrieslandCampina's 46. Of these five firms, two are cooperatives, and three are privately owned.

Table 1. Overview over analyzed firms

Strategy/Firms name	Country of origin	Organization	No. of countries with subsidiaries (2017)*	No. of brands (2017)	FSI (2017)
International strategy					
Ammerland	Germany	cooperative	5	1	40.8%
Bayernland	Germany	cooperative	2	2	43.0%
BMI	Germany	cooperative	3	6	44.2%
DMK	Germany	cooperative	10	11	43.1%
frischli	Germany	private	1	6	18.4%
Goldsteig	Germany	cooperative	2	2	32.9%
Käserei Champignon	Germany	private	5	10	40.0%
Omira	Germany	cooperative	1	6	37.2%
Rücker Aurich	Germany	private	2	1	47.7%
Rücker Wismar	Germany	private	1	1	0.0%
Uelzena	Germany	cooperative	1	8	19.5%
ø			**3.0**	**4.9**	**33.3%**
Global strategy					
Hochwald	Germany	cooperative	5	15	45.6%
Zott	Germany	private	10	16	56.5%
ø			**7.5**	**15.5**	**51.1%**
Multinational strategy					
Arla	Denmark	cooperative	44	30	75.3%
Ehrmann	Germany	private	8	16	49.7%
FrieslandCampina	Netherlands	cooperative	30	46	76.6%
Hochland	Germany	private	8	6	58.1%
Meggle	Germany	private	20	1	50.3%
ø			**22.0**	**19.8**	**62.0%**

* including Germany

Source: Authors' own calculation based on Ammerland, 2018; Bayernland, 2018; BMI, 2018; DMK, 2018; frischli, 2018; Goldsteig, 2018; Käserei Champignon, 2018; OMIRA, 2018; Rücker, 2018; Uelzena, 2018; Hochwald, 2018; Zott, 2018; Arla, 2018; Ehrmann, 2018; FrieslandCampina, 2018; Hochland, 2018; Meggle, 2018.

Two firms are not originally from Germany: FrieslandCampina is based in the Netherlands and Arla's home country is Denmark. It should be pointed out that these two firms are the only two cooperatives in the group employing a multinational strategy. The three firms headquartered in Germany that follow the Multinational Strategy – Ehrmann, Hochland and Meggle – are privately owned and based in the south of Germany. The average FSI in this group was 62%, with a range of 49.7% (Ehrmann) to 76.6 % (FrieslandCampina). Due to greater geographical distance and more complex firm structures, the management of these firms is nuanced – as can be seen for example at FrieslandCampina which, since 2016, has been managing its activities in China through a new business group "Consumer Products China" (FrieslandCampina 2016).

4.2 Empirical Results

Table 2, below, shows the minimum, maximum and geometric average values of the firms' EBIT Margin (EM), Turnover (T) and Gearing Ratio (GR) considered in this study. Some data were not available. Data for DMK over the years 2010 and 2011 are not available because DMK was only founded in 2012. For Bayernland eG, consolidated financial statements are only available from 2013 onwards. In the preceding years, only the financial accounting information of the individual firms which merged into Bayernland eG is available, but these are not comparable to consolidated financial statements. Data for Omira are only available until 2016, as the dairy was taken over by the French Lactalis Group in 2017.

The small group of firms that employ a global strategy had an average maximum EBIT margin of 3.8% (Zott: 6.6%; Hochwald: 1.0%). The average minimum value of this margin is 0.9%. During the observed period the average EBIT margin in this group was 2.7%. Zott's average EBIT margin was 4.7% —noticeably higher than that of Hochwald (0.8%). The minimum average EBIT margin in the group that use a multinational strategy is -0.3%. The lowest minimum values can be observed at Rücker Wismar (-5.2%), Omira (-2.2%) and frischli (-1.7%). The average maximum EBIT margin for the group of companies pursuing the International Strategy was 2.6%. OMIRA and Bayernland both posted the lowest EBIT margins of 1.2%, while the maximum EBIT margin was highest at Käserei Champignon (4.7%) and BMI (3.8%). Over the entire period under review, the average EBIT margin for this group was 1.4%, with Omira showing the lowest margin at 0.2% on average and Käserei Champignon the highest at 3.1%. In the group of companies with a multinational strategy, the minimum average EBIT margin was 1.9%. The figures fluctuated between -0.2% for Ehrmann and 2.9% for Arla. The maximum EBIT margin for this group averaged 6.3%. Hochland achieved the highest maximum EBIT margin at 8.7%, both in the group of companies with a multinational strategy and across all the groups considered. Over the entire period under review, the average EBIT margin of the companies in the Multinational Strategy group was 3.9%, with Hochland achieving the highest average EBIT margin at 4.9% and Ehrmann the lowest at 3.2%.

Table 2. Development of EBIT margin (EM), turnover (T) and gearing ratio (GR) for dairy firms with different internationalization strategies

Strategy/Firms name	EM (Min.)	EM (Max.)	EM ø2010–2017	T (Min., m. €)	T (Max., m. €)	T ø2010–2017	GR (Min.)	GR (Max.)	GR ø 2010–2017
International strategy									
Ammerland	0.6%	1.8%	1.2%	489.9	889.5	672.9	117.6%	159.4%	137.4%
Bayernland[1]	0.5%	1.2%	0.9%	647.3	786.7	699.7	191.0%	231.9%	210.0%
BMI	0.7%	3.8%	2.3%	420.1	622.8	535.3	157.7%	469.3%	241.9%
DMK[2]	1.1%	2.1%	1.5%	4,438.5	5,795.6	5,089.8	166.9%	221.1%	186.2%
frischli	-1.7%	3.5%	1.1%	379.1	573.4	449.2	164.1%	223.5%	184.1%
Goldsteig	1.0%	2.8%	2.2%	379.1	573.4	444.4	169.9%	264.9%	202.3%
Käserei Champignon	-0.2%	4.7%	3.1%	302.6	373.8	345.5	171.0%	210.5%	191.7%
Omira[3]	-2.2%	1.2%	0.2%	420.1	637.0	548.3	130.9%	299.7%	186.1%
Rücker Aurich	0.1%	1.3%	0.7%	302.7	446.4	383.2	405.6%	680.9%	526.2%
Rücker Wismar	-5.2%	3.7%	0.8%	132.1	214.0	183.6	345.2%	1790.8%	657.3%
Uelzena	1.6%	2.3%	1.9%	379.8	703.0	508.4	156.0%	289.0%	232.5%
ø	**-0.3%**	**2.6%**	**1.4%**	**766.4**	**1056.0**	**896.4**	**200.8%**	**440.1%**	**268.7%**
Global strategy									
Hochwald	0.5%	1.0%	0.8%	1,162.5	1,589.6	1,386.9	175.7%	264.1%	221.2%
Zott	1.4%	6.6%	4.7%	754.4	1,001.4	879.0	198.4%	317.8%	244.2%
ø	**0.9%**	**3.8%**	**2.7%**	**958.5**	**1295.5**	**1,133.0**	**187.1%**	**290.9%**	**232.7%**
Multinational strategy									
Arla	2.9%	5.2%	3.7%	6,577.6	10,614	9,133.9	171.1%	298.2%	237.9%
Ehrmann	-0.2%	7.2%	3.2%	582.6	767.1	710.6	79.0%	168.9%	122.0%
FrieslandCampina	2.7%	5.1%	4.3%	8,972	12,110	10,756	157.2%	185.5%	169.0%
Hochland	2.0%	8.7%	4.9%	1,055	1,446	1,203	47.1%	55.5%	50.8%
Meggle	1.8%	5.3%	3.5%	725.3	1,095.7	929.1	149.8%	182.7%	165.7%
ø	**1.9%**	**6.3%**	**3.9%**	**3582.5**	**5206.6**	**4,546.5**	**120.9%**	**178.2%**	**149.1%**

[1] 2013–2017 [2] 2013–2017 [3] 2010–2016

Source: Authors' own calculation based on Ammerland, 2018; Bayernland, 2018; BMI, 2018; DMK, 2018; frischli, 2018; Goldsteig, 2018; Käserei Champignon, 2018; OMIRA, 2018; Rücker, 2018; Uelzena, 2018; Hochwald, 2018; Zott, 2018; Arla, 2018; Ehrmann, 2018; FrieslandCampina, 2018; Hochland, 2018; Meggle, 2018

One of the control variables taken into account in this study is the firms' overall turnover. The average minimum turnover of companies pursuing an international strategy is € 766.4 million, with Rücker Wismar Dairy posting the lowest minimum turnover of € 132.1 million. The maximum average turnover of the companies considered in this group is € 1.06 billion, where DMK is by far the largest company in the group with a maximum turnover of € 5.8 billion. Over the period under review, the average turnover for all companies in the group was € 896.4 million.

In this group of companies, the minimum average turnover was € 958.5 million, with Hochwald's turnover of almost € 1.6 billion being higher than that of Zott (€ 754.4 million). The maximum average turnover in this group is € 1.3 billion, with Hochwald's turnover (€ 1.4 billion) being significantly higher than that of Zott (€ 1 billion). The average turnover over the entire period is € 1.13 billion.

With an average turnover of € 4.5 billion over the entire period under review, the turnover of companies with a multinational strategy was significantly higher than that of the other groups under consideration. At € 10.8 billion, Hochland has the highest average turnover, both in this group and across all the groups considered. At € 710.6 million, Ehrmann's average turnover is the lowest in this group. Ehrmann also has the minimum turnover in this group (€ 582.6 million) and FrieslandCampina the maximum (€ 12.1 billion).

With regard to companies' turnover, we find clear differences between the internationalization strategies pursued. The firms following the International Strategy show the lowest average turnover over the time period (€ 896.4 million) observed, followed by the firms employing the Global Strategy with an average turnover of € 1.1 billion over the observed period. The highest average turnover can be seen in the group of firms following the Multinational Strategy with an average turnover of €4.5 billion from 2010 to 2017.

Another control variable is the gearing ratio. For companies with an international strategy, the average gearing ratio over the period under review was 268.7%. At 137.4%, the Ammerland dairy had the lowest average EBIT margin and the lowest minimum gearing ratio in this group. At 657.3%, the Rücker Dairy in Wismar has the highest average gearing ratio in this group, as well as across all groups. The maximum value of 1790.8% is also the highest across all groups.

The companies pursuing a global strategy had an average gearing ratio of 232.7% over the period under review, with Hochwald (221.2%) and Zott (244.2%) being quite close to each other. Hochwald shows the minimum value in this group with 175.7%, Zott the maximum value in this group with 31.7%.

With an average gearing ratio of 149.1% over the observation period, the companies pursuing a multinational strategy achieved the lowest value of all comparison groups. With 50.8%, Hochland has the lowest average gearing ratio in this group, but also across all comparison groups. The minimum value of 47.1% is also the lowest across all companies considered. At 237.9%, Arla's average gearing ratio was highest in the group of companies with a multinational strategy. Moreover, Arla has the highest value in this group with 298.2%. The average minimum gearing ratio in this group is 120.9%, the maximum 178.2%.

Regarding the gearing ratios we find the highest values on average over the observed time in the group of firms following the International Strategy (268.7%), followed by the firms employing the Global Strategy (232.7%) and that employing the Multinational Strategy (149.1%).

4.3 Relationship between Internationalization Strategies and Economic Performance

Table 3 shows the correlation coefficients of the variables used in the model. As the results indicate there is not too strong a correlation between the coefficients to expect substantial problems with multicollinearity. This is also emphasized in the Variance Inflation Factor (VIF) statistic (see Table 2 in the Appendix). Further descriptive statistics on the variables can be seen in Table 1 in the Appendix.

Table 3. Correlation analysis

		1	2	3	4	5	6	7	8	9	10	11
1	EBITmargin	1										
2	Intern. Strategy	,520**	1									
3	Organization	,191*	,175*	1								
4	Turnover	,302**	,540**	-,388**	1							
5	GearingRatio	-,143	-,289**	,163	-,113	1						
6	DairyPriceIndex	-,078	,009	,013	-,004	-,022	1					
7	Fresh milk	-,040	-,156	-,506**	,409**	-,159	,011	1				
8	Fresh milk products	,004	,079	-,527**	,265**	-,329**	,041	,609**	1			
9	Cheese	,034	-,168*	-,389**	,158	,156	-,005	,016	-,211*	1		
10	Dried milk products	,004	-,190*	,038	-,145	-,135	,005	,108	,249**	-,219**	1	
11	Butter	-,128	,008	-,652**	,318**	-,006	-,012	,154	,332**	,438**	,299**	1

*p<0.05 **p<0.01
Source: Authors' own calculation

Table 4 shows the results of our random effects estimation. The results of our Hausman-Taylor estimation show a positive, highly significant influence of the multinational strategy on the economic success of firms analyzed in this study with a coefficient of 3.102 as shown in table 4. The gearing ratio has a positive significant effect on the firms' economic success. However, the coefficient of 0.001 shows that there is nearly no real impact on the firms economic performance due to a change in the gearing ratio. The production of cheese has a positive, highly significant influence on the EBIT-margin of the analyzed firms as the coefficient of 2.881 indicates. In addition, the production of dried milk products also has a positive effect on the economic performance of the firms surveyed, as can be seen from the significant coefficient of 1.528. Against this the production of butter significantly influences the firms' economic performance in a negative way as can be seen at the coefficient of -2.157.

Table 4. Results of the Hausman Taylor estimation

EBITmargin	**Coef.**		**Robust Std. Err.**	**P>z**
TVexogenous				
Turnover	0.000		0.001	0.921
Gearing Ratio	0.001	**	0.001	0.048
Dairy-Price-Index	-0.008		0.007	0.268
Fresh milk	0.014		0.374	0.971
Fresh milk products	0.528		0.594	0.374
TIexogenous				
Legal form	-0.142		0.825	0.864
Cheese	2.881	***	0.823	0.000
Dried milk products	1.528	***	0.456	0.001
Butter	-2.157	***	0.747	0.004
TIendogenous				
Intern. Strategy	3.102	***	0.539	0.000
Constant	0.149		1.449	0.918

*p<0.1 **p<0.05 ***p<0.01

Wald chi2(10) = 507.21; Prob > chi2 = 0.0000; sigma_u = 0.5739; sigma_e = 1.2511; rho = 0.1738

Std. Err. Adjusted for 18 clusters in ID; TV = time varying variables; TI = time invariant variables

Source: Authors' own calculation

Our analysis shows no significant influence of organizational form on the companies' economic success. Nor does the size of the company have any statistically significant influence on the economic success of the 18 companies we examined over the period

under review. The Dairy-Price-Index also has no statistically significant influence on the economic success of the companies.

We have checked for the robustness of the result in further calculations. Even if, for example, we exclude the internationalization strategy as a control variable, as it correlates relatively strongly with sales at 0.54, we found no significant influence for company size on economic success.

Due to multicollinearity problems with turnover, we were unable to examine the influence of the FSI and the number of brands on corporate success. As with sales, however, we found no real effect on corporate success in further calculations if we used these variables as control variables for turnover.

5 Discussion

The analysis shows clear positive effects of the Multinational Strategy on the economic performance of the firms considered compared to the International Strategy. Higher performance in this case also includes a risk premium for investing abroad (Busse and Hefeker, 2007). This result is supported by Heyder *et al.* (2011) and Qian (2002) studies, which showed the positive influence of internationalization on firms' economic performance. The results of Helpman *et al.* (2003) also underline the results of the study. According to the study, only the most productive companies serve foreign markets through foreign direct investments and local subsidiaries. Although the data basis does not allow us to measure productivity directly, we can assume that the higher productivity of the companies is reflected in the higher EBIT margins.

With regard to key figures, the Global Strategy is, as observed above, "stuck in the middle" between the International and Multinational Strategies. This can be explained by the Process Model of Internationalization described by Meißner and Gerber (1980). In this model, internationalization is seen as a multi-step process in which companies incrementally transfer capital and management from their country of origin to subsidiaries in foreign countries. Based on this theory, adopting a global strategy would be a step on the way from an international strategy to a multinational strategy, which is also reflected in key performance figures.

How can we explain the results and what do we learn from them? According to the theory of the management of multinational enterprises, a firm should use exports if doing so offers an advantage compared to its home country (Grant and Nippa, 2006); particularly as many export products, such as butter and milk powder, are undifferentiated and therefore compete through price margins which are likely to be low. Furthermore, more differentiated milk products, such as yoghurt, drinks and fresh cheese have limited export potential over great distances or have higher demands in regards to durability. Firms employing the Multinational Strategy are not limited by durability and thus seem better placed to fully exploit the potential of foreign markets. They can better adapt to local demand and conditions then build brands for local markets that will lead to higher turnovers (Harzing, 2000). We can see the highest number of brands at the firms employing the Multinational Strategy (see Table 1) – underlining this result. This is also underlined by the results of Qian (2002) who showed the positive interactive effects of multi-nationality and product diversification on the economic performance.

An aspect that limits the implementation of a multinational strategy is its factor demands. These, of course, include a higher capital demand compared to other strategies, especially the International Strategy, which is based mainly on exports. However, there are also soft factors such as management competencies as well as the availability of sufficient marketing, distribution and sales resources (Grant and Nippa, 2006; Theuvsen *et al.*, 2010).

Therefore, the sunk costs of foreign investments in subsidiaries are higher than those of simple exports, but the per unit costs are also lower than those of simple exports (Helpman *et al.*, 2003). It is therefore not surprising that the Multinational Strategy is employed by the (on average) larger and economically more successful firms considered in this study. We have countered this endogeneity problem with the Hausman-Taylor estimator and the possibility of estimating endogenous variables contained therein. However, with regard to capital demands, it is surprising that the companies employing the Multinational Strategy have the lowest average gearing ratios, in contrast to the other groups. Referring to the literature, one would expect higher gearing ratios in this group of dairies due to higher capital demands. Furthermore, the high number of brands within this group would give an expectation of a higher gearing ratio, compared to others, as there are higher production, marketing and legal costs for branding (Onkvisit and Shaw, 1989).

Interestingly, the price index for dairy products has no significant positive impact on the economic situation of dairy companies. This does not seem comprehensible at first, since rising product prices would, ceteris paribus, lead to the assumption of higher revenues and greater economic success. However, dairies pass on a substantial part of the higher prices to their farmers in the form of higher prices for raw milk. Furthermore, for many products, contracts are often concluded with customers for a certain period of time. Assuming that a company still has to service old contracts over a period of poor prices but at the same time has to increase the payout prices to its suppliers in order to keep up with companies in its neighborhood, this can have a negative impact on economic success. The same applies, of course, the other way around, when prices for dairy products are falling, when contracts from times of high prices are still being served but the payout price is already falling at the same time.

Compared to other studies, however, we found no significant influence of the different legal forms for companies' economic success. Other studies (Ebneth, 2006; Anderson and Henehan 2005; Jürgens *et al.*, 2015) point to the poorer economic performance of cooperatives compared to private companies and corporations. This is justified in the studies by corporate governance deficits (Ebneth, 2006; Anderson and Henehan, 2005) as well as the focus on less differentiated mass products with little added value (Jürgens *et al.*, 2015). On the other hand, we can confirm the results of Jürgens *et al.* (2015) to the effect that all German cooperative dairies, with the exception of Hochwald, focus on simple exports. The two cooperative dairies pursuing the Multinational Strategy are Arla and FrieslandCampina, which are the two cooperatives from abroad. This might be due to the small domestic market of these dairies. This alone does not explain why these dairies built up production plants worldwide and merged with foreign companies (Ebneth, 2006).

However, there are likely to be more factors that influence firms' economic performance. For example, product spectrum and competitive strategy have not been considered in this study and should be part of future analysis. In addition, the sample size is rather small due to non-availability of data. In the future, a larger data set in terms of observation time would allow analysis with other models such as a dynamic panel model. Indeed, the dairy companies headquartered in Germany considered in this study represent 60% of the total turnover of the German dairy industry and their foreign sales amount for 90.4% of the total foreign sales of the German dairy industry (Destatis,

2018). Future studies should be extended to other countries to check for the robustness of the results as far as data are available.

Nevertheless, internationalization and thus the right internationalization strategy will become more important. OECD and FAO forecasts predict 73% of future milk production growth in China and Asia up to 2025, while sales of fresh dairy products will increase versus concentrated products such as butter and cheese (OECD and FAO, 2016). This development can further favor the Multinational Strategy through localization – a fact that can be seen in the German dairy market's exports. Although the member states of the European Union are by far the most important customers for German dairy products across all dairy products, there are clear differences between the individual product groups. For example, exports of concentrated milk products to the European Union amounted to 67.3%, while exports of fresh products like buttermilk, curdled milk and yoghurt to the European Union accounted for 91.8% (Trademap, 2018).

In addition to these "hard facts", the implementation of a multinational strategy can offer further advantage. By producing "locally", the criticism of food exports from industrialized countries can be avoided, especially in many developing countries. A prerequisite, however, is an appropriate degree of professionalism in existing milk production and, if necessary, support from politicians and administrations in these countries

6 Conclusion

The study shows that there are significant differences regarding the influence of the different internationalization strategies on any firm's economic performance. The results show a positive influence of the Multinational Strategy on economic performance, compared to the other observed strategies. We can conclude therefore that dairy companies can gain economic advantages from localization and market adoption when internationalizing. Despite its higher requirements, especially in capital and management, the Multinational Strategy can pay off in the form of higher economic performance.

In the future, internationalization and thus the right internationalization strategy will become more important as OECD and FAO forecasts predict 73% of future milk production growth in China and Asia up to 2025, while sales of fresh dairy products will increase versus concentrated products such as butter and cheese (OECD and FAO, 2016). In addition, increasing requirements regarding animal husbandry and environmental protection lead to rising costs in domestic production. Although these additional costs on the domestic market can be partially offset by price premiums, e.g. for animal welfare standards, on the international, often price-oriented markets, this is by no means certain. Against this backdrop, German cooperative dairies in particular should review their internationalization strategy.

References

Ammerland (2011–2018). Annual Accounting of Ammerland 2010–2017. Annual accounting report of the Molkerei Ammerland e.G., German Bundesanzeiger, available at: https://www.bundesanzeiger.de/ebanzwww/wexsservlet (accessed 10 March 2019).

Anderson, B.L., Henehan, B. (2005). What gives agricultural co-operatives a bad name? *International Journal of Cooperative Management*, 2(2): 9–14.

Arla (2011–2018). Annual Report 2010–2017. Annual reports of Arla Foods amba, Viby J.

Bayernland (2011–2018). Annual Accounting of Bayernland 2010–2017. Annual accounting report of the Bayernland e.G., German Bundesanzeiger, available at: https://www.bundesanzeiger.de/ebanzwww/wexsservlet (accessed 10 March 2019).

Baltagi, B.H., Bresson, G., Pirotte, A. (2002). Fixed effects, random effects or Hausman-Taylor? A pretest estimator. *Economics Letters* 79: 361–369.

Becker-Blease, J.R., Kaen, F.R., Etebari, A., Baumann, H., (2010). Employees, Firm Size and Profitability of U.S. Manufacturing Industries. *Investment Management and Financial Innovation* 7(2): 7–23.

BMI – Bayerische Milchindustrie eG (2011–2017). Annual Accounting of the BMI 2010–2017. Annual accounting report of the Bayerische Milchindustrie e.G., German Bundesanzeiger, available at: https://www.bundesanzeiger.de/ebanzwww/wexsservlet (accessed 10 March 2019).

Busse, M., Hefeker, C. (2007). Political risk, institutions and foreign direct investments. *European Journal of Political Economy* 23: 397–415.

Capar, N., Kotabe, M. (2003). The Relationship between International Diversification and Performance in Service Firms. *Journal of International Business Studies* 34(4): 345–355.

Chaddad, F.R., Mondelli, M.P. (2013). Sources of Firm Performance – Differences in the US Food Industry. *Journal of Agricultural Economics* 64(2): 382–404.

Destatis 2018. Employees and turnover in the manufacturing industry. Genesis-Online database, Federal Office for Statistics.

DMK – Deutsches Milchkontor (2011–2018). Annual Accounting of the DMK 2010–2017. Annual accounting report of the Deutsche Milchkontor e.G., German Bundesanzeiger, available at: https://www.bundesanzeiger.de/ebanzwww/wexsservlet (accessed 10 March 2019).

Ebneth, O.J. (2006). Internationalisierung und Unternehmenserfolg, ein Vergleich europäischer Molkereigenossenschaften. *Schriften der Gesellschaft für Wirtschafts- und Sozialwissenschaften des Landbaues e.V.* 41: 363–374.

Ehrmann (2011–2018). Annual Accounting of Ehrmann 2010–2017. Annual accounting reports of the Ehrmann AG, German Bundesanzeiger, available at: https://www.bundesanzeiger.de/ebanzwww/wexsservlet (accessed 10 March 2019).

Fayerweather, J. (1989). *Begriff der internationalen Unternehmung.* Handwörterbuch-Export und internationale Unternehmung, Schäffer-Poeschel, Stuttgart: 926–948.

FrieslandCampina (2011–2018). Annual Report 2010–2017. Annual reports of Royal FrieslandCampina N.V., Armersfoort.

frischli (2011–2018). Annual Accounting of frischli 2010–2017. Annual accounting report of the frischli Milchwerke GmbH, German Bundesanzeiger, available at: https://www.bundesanzeiger.de/ebanzwww/wexsservlet (accessed 10 March 2019).

Goldsteig (2011–2018). Annual Accounting of Goldsteig 2010–2017. Annual accounting reports of the Goldsteig Käsereien Bayernwald GmbH, German Bundesanzeiger, available at: https://www.bundesanzeiger.de/ebanzwww/wexsservlet (accessed 10 March 2019).

Grant, R.M., Nippa, M. (2006). *Strategisches Management.* Pearson Studium, Hallbergmoos.

Guillouzo, R., Ruffio, P. (2005). Internationalization of European dairy co-operatives. *International Journal of Co-operative Management* 2(2): 25–35.

Harzing, A.W. (2000). An empirical analysis and extension of the Bartlett and Ghoshal typology of multinational companies. *Journal of International Business Studies* 31(1): 101–120.

Helpman, E., Melitz, M.J., Yeaple, S.R. (2003). Export versus FDI. Working Paper 9439, National Bureau of Economic Research.

Heyder, M., Makus, C., Theuvsen, L. (2011). Internationalization and Firm Performance in Agribusiness: Empirical Evidence from European Cooperatives. *International Journal on Food System Dynamics* 2(1): 77–93.

Hochland 2011–2018. Annual Accounting of Hochland (2010–2017). Annual accounting report of the Hochland SE, German Bundesanzeiger, available at: https://www.bundesanzeiger.de/ebanzwww/wexsservlet (accessed 10 March 2019).

Hochwald (2011–2018). Annual Accounting of Hochwald 2010–2017. Annual accounting report of the Hochwald Milch e.G., available at https://www.bundesanzeiger.de/ebanzwww/wexsservlet (accessed 10 March 2019).

Holtbrügge, D., Welge, M.K. (2015). *Internationales Management – Theorien, Funtionen, Fallstudien*. Schäfer-Poeschel.

Hurdle, G.J. (1974). Leverage risk, market structure and profitability. *The Revue of Economics and Statistics* 54(4): 478–485.

Johnson, G., Whittington, R., Andwin, D., Regner, P. (2016). *Strategisches Management – eine Einführung*. Pearson Studium.

Jürgens, K., Fink-Keßler, A., Poppinga, O., Wohlgemuth, M. (2015). Wertschöpfung von Molkereien. MEG Milch Board, Göttingen.

Käserei Champignon (2011–2018). Annual Accounting of Käserei Champignon 2010–2017. Annual accounting report of the Käserei Champignon Hofmeister GmbH & Co. Kg, German Bundesanzeiger, available at: https://www.bundesanzeiger.de/ebanzwww/wexsservlet (accessed 10 March 2019).

Li, L. (2007). Multinationality and performance: A synthetic review and research agenda. *International Journal of Management Reviews* 9(2): 117–139.

Magaziner, I.C., Reich, R.B. (1985). International Strategies. *Strategic Management of Multitinational Corporations: The Essentials*. John Wiley & Sons: 4–9.

Meggle (2011–2018). Annual Accounting of Meggle 2010–2017. Annual accounting of the Meggle AG, German Bundesanzeiger, available at: https://www.bundesanzeiger.de/ebanzwww/wexsservlet (accessed 10 March 2019).

Meißner, H.G., Gerber, S. (1980). Die Auslandsinvestitionen als Entscheidungsproblem. *Betriebswirtschaftliche Forschung und Praxis*: 217–228.

MIV (2018). Wohin die Milch fließt. Dairy Industry Association of Germany, available at: https://milchindustrie.de/wp-content/uploads/2018/11/Wohin-die-Milch-flie%C3%9Ft-2018.pdf (accessed 11 October 2018).

OECD, FAO (2016). OECD-FAO Agricultural Outlook 2016–2025. Paris, OECD Publishing, Paris.

Onkvist, S., Shaw, J.J. (1989). The International Dimension of Branding: Strategic Considerations and Decisions. *International Marketing Review* 6(3): 22–36.

OMIRA (2011–2018). Annual Accounting of OMIRA 2010–2017. Annual accounting report of the OMIRA Oberland Milchverwertung Gesellschaft mit beschränkter Haftung, German Bundesanzeiger, available at: https://www.bundesanzeiger.de/ebanzwww/wexsservlet (accessed 10 March 2019).

Negandhi, A.R., Welge, M. (1984). Beyond theory. Global rationalization strategies of American, German and Japanese multinational companies. *Advances in International Comparative Management*, JAI Press: 193–201.

Perlmutter, H.V. (1969). The Tortuous Evolution of the Multinational Corporation. *Columbia Journal of World Business* 4: 9–18.

Qian, G. (2002). Multinationality, product diversification and profitability of emerging US small- and medium-sized enterprises. *Journal of Business Venturing* 17: 611–633.

Rücker Aurich (2011–2018). Annual Accounting of Rücker 2010–2017. Annual accounting report of the Rücker GmbH, German Bundesanzeiger, available at: https://www.bundesanzeiger.de/ebanzwww/wexsservlet (accessed 10 March 2019).

Rücker Wismar (2011–2018). Annual Accounting of Ostsee-Molkerei Wismar 2010–2017. Annual accounting report of the Ostsee-Molkerei Wismar GmbH, German Bundesanzeiger, available at: https://www.bundesanzeiger.de/ebanzwww/wexsservlet (accessed 10 March 2019).

Scholl, R.F. (1989). *Internationalisierungsstrategien.* Handwörterbuch Export und internationale Unternehmung, Schäffer-Poeschel: 983–1001.

Theuvsen, L., Ebneth, O. (2005). Internationalization of Cooperatives in the Agribusiness – Concepts of Measurement and their Application. In: Theurl, T. (Ed.), *Strategies for Cooperation*, Shaker, Aachen: 395–419.

Theuvsen, L., Janze, C., Heyder, M. (2010). Agribusiness in Germany 2010. On the Way to New Markets, Ernst & Young, Stuttgart.

Thomas, D.E., Eden, L. (2004). What is the shape of the multinationality-performance relationship? *Multinational Business Review* 12(1): 89–110.

Trademap (2018). List of importing markets of product groups 0401, 0402, 0403, 0404, 0405, 0406 exported by Germany. Inetrnational Trade Center, available at: https://trademap.org/Index.aspx (accessed 20 December 2018).

Uelzena (2011–2018). Annual Accounting of Uelzena 2010–2017. Annual accounting report of the Uelzena e.G., German Bundesanzeiger, available at: https://www.bundesanzeiger.de/ebanzwww/wexsservlet (accessed 10 March 2019).

Verbeek, M. (2004). *A Guide to Modern Econometrics.* John Wiley & Sons Ltd.

Vermeulen, F., Barkema, H. (2002). Pace, Rhythm, and Scope: Process Dependence in Building a Profitable Multinational Corporation. *Strategic Management Journal* 23: 637–653.

Vitaliano, P. (2016). Global Dairy Trade: Where Are We, How Did We Get Here and Where Are We Going?. *International Food and Agribusiness Management Review* 19: 27–36.

Wind, Y., Douglas, S.P., Perlmutter, H.V. (1973). Guideline for Developing International Strategies. *Journal of Marketing* 37(2): 14–23.

Zott (2011–2018). Annual Accounting of Zott 2010–2017. Annual accounting report of the Zott SE & Co. KG, German Bundesanzeiger, available at: https://www.bundesanzeiger.de/ebanzwww/wexsservlet (accessed 10 March 2019)

Appendix

Table 1. Descriptive statistics

Variable		Mean	Std. Dev.	Min	Max	Observations
ID	overall	9.54	5.28	1	18	N = 138
	between		5.34	1	18	n = 18
	within		0	9.54	9.54	T-bar = 7.67
Year	overall	4.57	2.27	1	8	N = 138
	between		0.44	4	6	n = 18
	within		2.24	1.07	8.07	T-bar = 7.67
EBIT-margin (%)	overall	2.33	1.98	-5.2	8.7	N = 138
	between		1.50	0.2	4.8625	n = 18
	within		1.323685	-3.70	6.28	T-bar = 7.67
Internationalization-Strategy (dummy)	overall	0.29	0.46	0	1	N = 138
	between		0.46	0	1	n = 18
	within		0.00	0.29	0.29	T-bar = 7.67
Organization (dummy)	overall	0.46	0.50	0	1	N = 138
	between		0.51	0	1	n = 18
	within		0	0.46	0.46	T-bar = 7.67
Turnover (million Euros)	overall	1927.85	3107.64	132.08	12110	N = 138
	between		3120.33	183.58	10756.13	n = 18
	within		429.95	-628.48	3407.93	T-bar = 7.67
Gearing Ratio (%)	overall	229.82	180.05	47.1	1790.80	N = 137
	between		141.32	50.85	657.27	n = 18
	within		117.64	-82.25	1363.35	T-bar = 7.67
Dairy-Price-Index (index value)	overall	201.18	30.24	153.77	242.75	N = 138
	between		1.69	196.13	201.60	n = 18
	within		30.21	153.35	247.80	T-bar = 7.67
Frischmilch (dummy)	overall	0.52	0.50	0	1	N = 138
	between		0.50	0	1	n = 18
	within		0.12	0.15	1.15	T-bar = 7.67
Frischmilchprodukte (dummy)	overall	0.75	0.44	0	1	N = 138
	between		0.43	0	1	n = 18
	within		0.12	0.12	1.12	T-bar = 7.67
Käse (dummy)	overall	0.88	0.32	0	1	N = 138
	between		0.32	0	1	n = 18
	within		0	0.88	0.88	T-bar = 7.67
Trockenprodukte (dummy)	overall	0.73	0.44	0	1	N = 138
	between		0.46	0	1	n = 18
	within		0.00	0.73	0.73	T-bar = 7.67
Butter (dummy)	overall	0.59	0.49	0	1	N = 138
	between		0.50	0	1	n = 18
	within		0.00	0.59	0.59	T-bar = 7.67

Source: Authors' own calculation

Table 2. VIF statistics

Variable	**VIF**	**1/VIF**
Legal form	4.91	0.204
Intern. Strategy	3.78	0.264
Butter	3.71	0.269
Turnover	3.38	0.296
Fresh milk products	3.00	0.334
Fresh milk	2.88	0.347
Dried milk products	2.12	0.473
Cheese	2.04	0.491
Gearing Ratio	1.53	0.654
Dairy-Price-Index	1.01	0.994
Mean VIF	**2.83**	

Source: Authors' own calculation

VI Prospects and Ways for the Supply Chain Integration in the Dairy Sectors of Brazil and Germany

Authors: Johannes Meyer, Caetano Beber

Submitted to: Agricultural Economics

Abstract

Today, Brazil is a competitive player on international markets for many agricultural products. However, the same does not apply to the country's dairy industry. On the other hand, the German dairy industry faces a saturated domestic market, an increasing liberalization of European agriculture and trade policy, as well as domestic political and social demands. Despite internal resistances, it is slowly orienting itself towards international competition. Therefore this paper investigates prospects and ways for an integration the Brazilian and German dairy supply chains and how they can benefit from this situation. By using expert interviews in both countries and firm level data of German dairies, this paper offers an holistic overview to identify risks and opportunities of a possible integration of the supply chains. Finally, ways of dealing with potential barriers and shortcomings are proposed. The analyses of this paper show that the problems of the industries in both countries basically complement each other very well at different levels. There are positive synergies for both sides when it comes to positioning themselves competitively for the future for conquering market shares in a globally growing milk market. By bridging development economics and agribusiness research this paper proposes an innovative framework of assessment for the challenges of such integration, as well as viable solutions to globalization issues.

1 Introduction

The globalization of markets and the internationalization of economic activities is likely to be an irreversible path that in past decades raised international competitive pressures on agri-food supply chains of both developed and developing countries. The consequent modernization of these supply chains, in several cases led by processing companies, might represent a strategic opportunity for developing countries, where they can not only provide new sources of FDI, revenues and employments but also access to technology and knowledge (Wilkinson, 2004). From the developed countries' perspective, it creates opportunities for accessing new resources and for market expansion in light of their market saturation at home. On the other hand, the literature also shows that the absence of regulation in global value chains dominated by multinationals pursuing only outsourcing strategies, generally doesn't generate the benefits otherwise possible. In such cases companies act to take advantage from weak institutions, low wages and less strict environmental laws in developing countries, simply to expand their power and market shares with no or very little compensation to their hosts (Reardon *et al.*, 2009).

We assume that the supposed benefits are more likely to be achieved when the internationalization initiative comes from an intermediary level in the supply chain, the dairy industry. These are the stakeholders that maintain the most direct contact with up and downstream elements, have the power to mediate in conflicts and can maintain a fair power balance along the entire chain (Tybout, 2000; Friedrich, 2010). We therefore focus our analyzes on the dairy industry in a developed country, Germany, and a developing/emerging economy, Brazil , with the aim of identifying the constraints and success factors of the dairy industries in both countries and ways to raise them. As the dairy industry is always strongly linked to milk production due to perishability and high transport frequency for raw milk (Friedrich, 2010), we consider this in our analysis in both countries as well.

Brazil is a strong player in the agri-food business scenario. Brazilian figures amongst the most competitive players in today's international markets in agri-food sectors such as soybean, corn, sugar cane, pork and poultry. On the other hand, there is a non-competitive dairy industry that has not benefited from this change or has benefited only marginally from it (Helfand *et al.*, 2015; Beber *et al.*, 2019). A completely different situation can be found on the other side of the Atlantic in Germany. The German dairy industry is among the biggest exporters worldwide. However, in the future, it faces certain challenges which will make it more difficult to gain and maintain market share because, on the one hand, resources for milk production such as arable land are scarce and milk production is in competition with other types of production with better opportunity costs. On the other hand, societal requirements on production conditions related to the environment and animal welfare (e.g. GMO free feeding) are increasing. That means that under certain circumstances the additional costs may be recovered on the domestic market for specialty products but that certainly does not apply to mass-produced goods on the international markets (BVE, 2017).

With the combination of qualitative and quantitative methods, this study investigates whether it is possible to conceive a framework where the internationalization of a supply chain generates a win-win situation for both hosts and guests. A process that is context-adapted and involves local partners to mitigate eventual cultural shocks. An integration that has the capacity to jointly reduce weaknesses and boost strengths, increasing overall competitiveness to achieve individual objectives and interests.

This paper contributes to the literature on the internationalization of food supply chains, taking into account socio-economic and cultural differences to search for smoother ways of integration North-South. It also proposes an innovative framework for assessing the challenges of such integration as well as viable solutions to globalization issues. In doing so it bridges development economics and agribusiness research, where the sustainable agro-industrialization process of this supply chain might contribute to competitiveness, business development and poverty alleviation (Cook and Chaddad, 2000; Reardon and Barrett, 2000).

In the next section we detail the structure and the development of the dairy sector in both countries. Next, after an extensive survey of dairy industry leaders in both countries, we analyze opportunities and risks for both sides arising from the given situation. In sequence we analyzed the German dairy processing companies' current internationalization strategies using the Perlmutter's EPRG Model and the success factors which impact on their performances. Finally we derived possible combined solutions for the Brazilian and the German dairy supply chains, which in the best case represent a win-win situation for both sides.

2 Structure and Development of the Dairy Sector in Brazil and Germany

2.1 Milk Production

When considering the dairy industry in both countries, it should first be noted that the dairy industry and associated industries are not evenly distributed across the country but are instead concentrated in certain regions and also differ in some aspects of their production intensity. In the case of Brazil, milk production has grown since the 1990s mainly due to product increases in the Southern[1] region, especially in the "Grande Fronteira do Mercosul" area. Production in this area is mainly pasture-based (Gonçalves de Assis *et al.*, 2005), based on family farms and cooperative enterprises (Anschau, 2011). Whereas in 2006 the share of the South in total Brazilian milk production of 25.4 billion liter was 13.3%, by 2015 it had risen to 18.5%, with a total national production of 35 billion liter (IBGE, 2017). Against this background, the South of Brazil is part of our investigation.

Milk production in Germany is mainly distributed among regions where fewer types of land use than grassland are available, resulting in low opportunity costs for the land factor (Görmann *et al.*, 2006). Accordingly, milk production is concentrated around the North Sea coast, in the low mountain ranges of central Germany, in the Alps and along the Czech border. Exceptions to this are the Lower Rhine region and the region along the Dutch border (Lassen *et al.*, 2009). As in Brazil, milk production is continuing to intensify in some regions. The share of intensive regions in Germany's total milk production rose from 58.8% in 2010 to 59.8% in 2015. Looking at production growth, the importance of the intensive regions becomes even clearer: in 2015 their share of production growth was 78.5% (Meyer and Theuvsen, 2017).

In production, Brazil also presents extremely low productivity in dairy farming, with less than 1963 kg/animal/year, while New Zealand for example, with a similar pasture-based system and pedo-climatic conditions as Brazil, produces approximately 4237 kg/animal/year. The neighboring countries of Argentina and Uruguay also show higher productivities, 3001 and 2645 kg/animal/year respectively in 2017 (FAOSTAT, 2019b). However, in recent years these numbers have been changing in Brazil. A restructuring process in dairy production is taking place, generating notable improvements.

[1] The South of Brazil comprises three states: Paraná, Santa Catharina and Rio Grande do Sul

In 1996 there were 1.81 million farmers in Brazil involved in dairy production; of these only 1.35 million still remained active by 2006 and only 1.17 million in 2017, showing an exit rate of 35% among dairy producers within only 21 years (IBGE, 1996, 2006, 2017). Meanwhile total production increased by 81% over the same period to 33.5 billion liters in 2017 (see Figure 1). These figures highlight gains in scale and productivity at the national level. This is also indicated by sector statistics from the five largest producer states in Brazil (Mina Gerais, Rio Grande do Sul, Santa Catarina, Paraná and Goiás), which correspond to 71% of the total national production. Thereby, the collective productivity average of these regions in 2014 was 1968 liters/cow/year and thus higher than the country average of 1525 liters/cow/year (IBGE, 2016). The national herd increased until 2014, then decreased to around 17.1 million animals in 2017. In this period the productivity per cow increased by 29% (IBGE, 2018) led by Mina Gerais (62%) and Santa Catarina (33%), the most productive state in the country with 3,580 liters/cow/year. These recent developments highlight the structural changes taking place in Brazil and a higher technification of the production systems reflecting efforts to achieve gains in efficiency and productivity.

Despite the recent improvements, the last complete agricultural census (IBGE, 2006) presents parameters that show the reality of low competitiveness and technology adoption in the dairy production at that time. For instance, 'mechanical milking' was present on only 2.4% of farms representing 22% of the milk collected. Artificial insemination was present on 1.4% of farms representing 14% of the milk. Finally, only 11% of farms had cooling tanks. Small farms predominate; 84% own less than 50ha corresponding to 60% of the total production quantity and 45% produced less than 10 liters/day.

This is probably one of the reasons why the cost per kg of milk in Brazil, despite the very good natural conditions, is still high and milk production less competitive. Production costs in neighboring countries directly south, Argentina and Uruguay, are 27 and 29.5 US-$ per 100 kg energy corrected milk (ECM) for medium-sized dairy farms and 25.5 and 28 US-$ per 100 kg ECM for large dairy farms. On the other hand, the production costs on both the average large farm and medium Brazilian dairy farms are around US$ 40 per 100 kg ECM (IFCN, 2013).

In contrast, German milk production has been influenced by European agricultural policy since advent of the European Economic Union. Due to the strong promotion of production and protectionist measures, production soon exceeded demand, so that the “milk quota” was introduced in 1984. At the end of the 1980s, this led to a reduction in intervention stocks and expenditure on the milk market organization. At the end of 2015 the milk quota was abolished in the European Union and thus also in Germany. The abolition was preceded by an increase in the milk quota from 2007 onwards (Kreß, 2005; BMEL, 2018). These framework conditions are also reflected in the development of milk production in Figure 1. It is clear that German milk production started to rise again from 2006, after years of stagnation. As a result of the increase in production, export surpluses also rose, as the German market was already saturated. The high export surpluses before 2006 indicate that Germany had been an important exporter on the milk market in previous years. From 2014, however, increases in milk production would decline again (see Figure 1).

Figure 1. Development of milk production and trade balances in Brazil and Germany

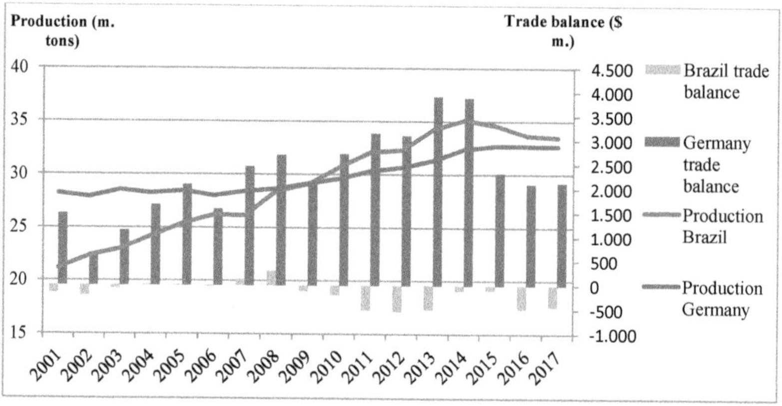

Source: Authors’ own depiction and calculation based on FAOSTAT, 2019a; Trademap, 2019.

2.2 Dairy Trade and Industry

As Table 1 shows, the two countries differ considerably in structure, both in agricultural milk production and in the structure of the dairy processing industry. The low technology adoption rates in Brazilian milk production are also connected to low quality in dairy outputs. In this regard, Brazil recently obtained an international sanitary certification to export to China only in 2015. In 2017 around 30% of Brazil’s total milk production was self-consumed or traded on informal markets. The remaining 70% was processed by about 1,930 companies (IBGE, 2018).

Table 1. Overview of key figures for the dairy industry in Brazil and Germany in 2017

Category	Brazil	Germany
Arable land (m. ha)	77.3	11.8
Grassland (m. ha)	158.6	4.7
Milk produced (m. tons)	33.5	32.6
Number of dairy farms	1,171,190	65,782
Number of cows (m.)	17.1	4.2
Number of cows per farm	15	64
Average yield per cow (kg/year)	1,960.0	7,763.0
Ratio of produced and delivered milk (%)	72.0*	96.4*
Dairy imports (€ m.)	545.4	7,570.1
Dairy exports (€ m.)	102.2	9,693.6
Turnover of milk processing industry (€ bn.)	19.5	27.9
Number of milk processing companies	1,930	219
Average turnover of milk processing companies (€ m.)	10.1	127.2

* 2015

Source: Authors' depiction based on BLE, 2019; FAOSTAT, 2019a; Destatis 2019a, 2019b, 2019c; Trademap, 2019;IBGE, 2018; Carvalho *et al.*, 2018.

In Brazil the national industry suffers from import pressure from neighboring countries such as Uruguay and Argentina, countries with more solid dairy industries. Those imports arrive in the country at lower prices, despite import tariffs, emphasizing the low competitiveness of the national processing industry. The country shows an historical negative dairy trade balance. Policy support is sporadic and mostly via subsidized short and long-term loans from federal rural credit systems (Chaddad and Jank, 2006). As in most developing and emerging countries, the agricultural sector, which produces around one third of Brazil's GDP, is important in sustaining the country's industrial growth and is therefore more heavily taxed and less subsidized (Khan, 2001; Stiglitz, 1987). Dairy supply chains have experienced significant concentration at all levels in most OECD countries (OECD, 2001). This also happens in the southern region of Brazil, where from the 469 processing companies existing in 2012, only 383 (-18%) still remained in 2017, while volumes collected remained stable at around 8.6 billion liters per year. Efficiency gains and countervailing market power arguments have been offered as explanations for increases in concentration at various stages of the dairy supply chain.

An indicator that reflects the development and professionalization of the two production systems at national level is the ratio of milk delivered to milk produced. While in Germany, 96.4% of all milk produced is delivered to dairies, in Brazil only 72.10% of the milk produced reaches the dairies. Other examples such as New Zealand (99.9%), the USA (99.5%) and even Argentina (93%) have higher shares of milk industrialized by dairies (ZMB, 2016).

The German dairy industry has significantly increased its milk production in recent years as a result of the abolition of the milk quota. Accordingly, the dairy industry's turnover increased to €27.9 billion in 2017. Against the background of a saturated domestic market, the German dairy industry's export quota rose from 27.8% in 2008 to 33.4% in 2016, reaching its highest level in 2014 with an export share of 34.2% of total turnover (Destatis, 2019a, 2019b). In terms of physical volume, the importance of exports is much more important. At 49.1%, just under half of the milk processed in Germany is exported (MIV, 2017). Approximately 50 to 60% of the collected milk is processed in cooperative dairies (Hörl and Hess, 2017).

With regard to foreign trade in dairy products, it is striking that cheese, concentrated dairy products and cream are the German dairy industry's largest export products in terms of turnover. On the other hand, concentrated milk products and buttermilk account for 92.1% of the positive trade balance. In the case of Brazil, concentrated milk products and cheese account for the lion's share (88.6%) of the negative trade balance. Brazil only has a positive trade balance in fresh milk and cream (Trademap, 2019).

However, when considering worldwide trade in dairy products, the geographical aspect and the product group should also be taken into account in addition to sales and volumes. In 2016, 84.2% of Germany's exports by value were exported to EU member states. At 91.8%, the share of exports to the European Union was highest for fresh milk products. It was lowest for concentrated milk products at 67.1%. This aspect also applies to most major exporters of dairy products such as the USA and New Zealand. Mexico, for example, is the most important buyer of milk products from the USA, and Brazil the most important from Argentina. An exception to this is New Zealand, whose most important partner is China. Its exports are mainly concentrated dairy products such as milk powder (Trademap, 2019).

2.3 Why Brazil?

A recent report in the Dairy Reporter (Riley, 2017) draws attention to the fast growing Latin American (LA) dairy market. While companies look to Asia – and China in particular – to expand and provide a lucrative market for dairy products, there is another market that has continuously shown solid growth and opportunity. Latin America is a region where dairy companies have been able to introduce new products to an eager consumer base and capitalize on an established market. Despite being one of the fastest growing in the world, at US$60bn the LA dairy market accounts for less than 15% of global dairy sales. The economic downturn in several markets and a weak demand have been reflected in dairy sales, but the continued urbanization of LA cities (80% of LA citizens now live in urban areas) is pushing dairy forward.

In this context Brazil stands out, boasting a dynamic dairy sector, which offers significant opportunities for growth and development. Traditionally Brazil is a strong player in the global agri-food business scenario. It is therefore an important case to be analyzed due to the large consumer market and to comparative and competitive advantages regarding total production. From the demand side it accounts for 207 million inhabitants, and US$ 1.8 trillion gross domestic product (GDP) in 2016 (US$ 2.62 in 2010). In recent years, internal demand for dairy products has increased significantly thanks to the growing purchasing power of Brazilian consumers. Over the past 10 years, middle-class consumers have grown from 38% to 56% of the population and today include more than 119 million people (IBGE, 2018).

On the supply side, the abundance of resources is remarkable. With 8.5 million km2, nearly 19% of world's arable land is in Brazil (OECD and FAO, 2015), where only 10% is being used. Moreover, 19% of the planet's fresh water is in Brazil. With a wide range of latitudes and reasonably well-distributed rainfall throughout the year, the country is able to produce a wide range of products all year round. Consequently, agricultural production and productivity both have large growth potential through the incorporation of new areas and technology adoption (Farina and Nunes, 2002). It is not surprising that this country has attracted a considerable amount of foreign direct investment (FDI) in the food sector and especially in the dairy over the past several years. We can pinpoint the following items as being main factors in attracting foreign capital to Brazil: a) dimension of the Brazilian market; b) interest in making Brazil an export base to the

Mercosur (and Latin American) trade partners; c) economic stability; d) fiscal incentives; e) access to raw materials; f) low cost of labor (Farina and Viegas, 2003).

Attentive to such opportunities, big conglomerates such as Nestlé and Lactalis have recently acquired several large processing plants in GFM. For example Lactalis only arrived in the GFM in 2014, and by 2015 it was already the largest group in the dairy sector in GFM (Beber *et al.*, 2019) with a collection capacity of around 2.7 million liters/day.

The cultural aspects may also play in favor of adapting the German processing companies in the South of Brazil. They might enjoy lower cultural barriers and faster product adoption because of the large number of German descendants in that area. German colonization in the South of Brazil officially started in 1824. Its objective was to consolidate possession and maintenance of territory through settlement. Since then, German descendants have played an important role in that area, contributing to the diversification of agriculture, urbanization and industrialization, also spreading their influence into many areas, from architecture to religion, food and culture. Over the past two centuries, more than 250 thousand Germans arrived in Brazil. In the 1930s, 20% of the population of the Southern states was already of German origin. Currently, it is estimated that 10% of the Brazilian population have at least one German ancestor. In 1996, Damke (2008) estimated there were more than two million speakers of a variety of German dialects in Brazil, mostly in the South.

In light of such opportunities, but also considering several issues that must be resolved in the GFM, as illustrated in Chapter 2, its dairy supply chain requires the savoir-faire of experienced players in order to make it a strong competitor in the Global dairy scenario.

3 Methodology and Data

3.1 Qualitative Analysis

The research was conducted in the main producing zones in both Brazil and Germany, where qualitative primary data was collected concerning the competitive advantages and disadvantages of each supply chain. For the former, data was collected between November 2016 and January 2017 in the Southern Region, in three states that form the

'mesoregion Grande Fronteira do Mercosul[2] (GFM)'. For the later, data was collected between April and August 2018 in the two main dairy producer states, Lower Saxony and Bavaria.

A total of sixty-four interviews were conducted across a spectrum of leadership roles in the dairy industry; twenty-six in Brazil and thirty-eight in Germany (see Table 2). We interviewed managers, directors and presidents of the main dairy processing cooperatives and private companies (all large and medium enterprises), in addition to the leadership of institutes, associations, unions and producer organizations involved in the dairy sector. We intended to collect opinions from different perspectives in this economic activity. We selected the interviewees according to the snowball sampling method. This constitutes a non-probabilistic sample, used to study complex phenomena. A first sub-group of the population is interviewed, which itself then identifies other members of the group, who, in turn, refer to further people belonging to the same group, and so on (Goodman, 1961).

Table 2. Interviewed stakeholders in Brazil and Germany

Country	**Farmers**	**Political Association/Union**	**Producers Organization**	**Upstream; R&D; advice; supply**	**Downstream; processing; retail**	**Total**
Germany	10	8	3	12	5	38
Brazil		9	2	5	10	26
TOTAL	**10**	**17**	**5**	**17**	**15**	**64**

Source: Authors' own depiction

The actors were separated according to their activities in the supply chain (similar issues and support for competitiveness). They are:

1) Farms (FA)
2) Political (association / union / government) (GOV)
3) Producers organization (PO)
4) Upstream actors (research and consulting, services, extension) (US)
5) Downstream actors (processors, retailers) (DS)

We chose the main companies with operations in the region and the main institutes carrying out important actions to promote the supply chain. Some of the companies or cooperatives interviewed are the largest in the zone, representing in some cases more than

[2] The mesoregion comprises the Southwest zone of Parana, the West of Santa Catarina, and the Northwest of Rio Grande do Sul.

6,000 producers in GFM or more than 8,000 in Lower Saxony for example, and covering areas in more than one state. When considering the subsidiaries, associations and alliances, they are all on a larger scale and these organizations are usually dispersed all over the country. For confidentiality purposes, the interviewees are identified throughout the results by their acronym ("PO" for producer organization for example) followed by the country acronym (BR for Brazil and GE for Germany), then followed by an identification number. For example the Producer Organization 02 in Germany is identified as "POGE02".

Farmers were only interviewed in Germany. In Brazil they were not interviewed because of the high heterogeneity of production systems throughout the zone and limitations on data collection. We chose the German farms based on their competitive strategies. Different types of strategies were selected as long as they remain competitive / stable in the market. Farms were selected with the help of actors upstream, i.e. research and consulting institutions, and service providers, which are in direct contact with the producers.

Data was collected using semi-structured interviews, which were individually prepared and guided to avoid missing important aspects from each respondent. Questions varied according to the participants targeted. They covered aspects of the background information of those interviewed and their relation/influence on the supply chain with an historical perspective; structural and organizational aspects; management aspects; governance environment; market dynamics and external factors; technology adoption and diffusion; attributes of purchased raw milk; product differentiation and commercialization channels; future expectations and actions. The intention was to capture the main problems and strategy factors that might have any effect on sources of competitiveness and coordination between actors in the supply chain. Ten interviews were conducted in Rio Grande do Sul, eight in Santa Catarina and eight in the state of Paraná, showing a uniform spatial distribution within the GFM zone. Similar action was taken in Germany, with seventeen interviews conducted in Lower Saxony, sixteen in Bavaria and five at the national level. Each interview lasted around one and a half hours on average. In a few cases more than one person from the same institute or enterprise were interviewed.

After collection, the information was transcribed and a discourse/content analysis of the qualitative data was carried out including codification, a first round of analysis and recodification. We identified how the elements are related to each other and how they affect the competitiveness of each zone. From this process and from the fundamental topics investigated in this study, the competitive advantages and disadvantages emerged (according to the interviewees' perceptions). As a result, several factors were identified that directly or indirectly affect competitiveness in these supply chains and we will discuss them in order to answer the question of whether there can be a win-win situation for both countries in an internationalization process.

3.2 EPRG Model and Impact of Different Internationalization Strategies

Beside the qualitative analysis we also refer to quantitative data in the study of Meyer *et al.* 2019b. They analyzed the effect of different internationalization strategies on the economic success of 16 German dairies, one Dutch dairy and one Danish/Swedish dairy based on the companies' annual financial reports over the years 2010 to 2017. The different internationalization strategies were identified with the expanded Perlmutter's EPRG model (1969). This differentiates between internationalization strategies with regard to their advantages through globalization and standardization and through localization and differentiation (Wind *et al.*,1973) (see Figure 2).

Figure 2. EPRG Model

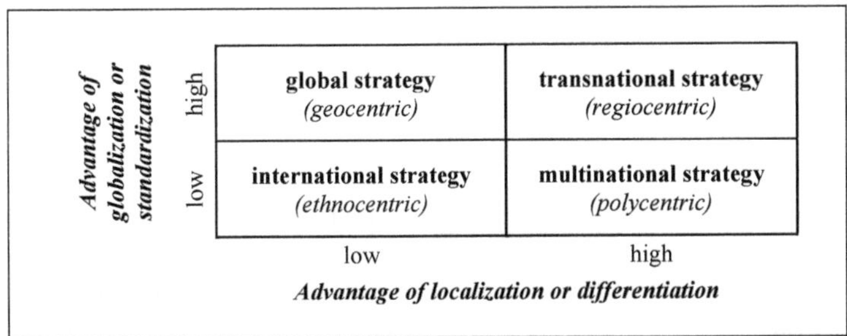

Source: Authors' own depiction based on Holtbrügge and Welge, 2015.

The multinational strategy is characterized by the fact that the companies that follow it have production sites abroad and also process milk there. These plants are located in more distant countries and partly on other continents. In contrast, the companies that follow the global strategy also have branches abroad, but these are located in Germany's neighbouring countries. The companies that follow the international strategy, process milk exclusively in Germany; foreign companies are only used for marketing and distribution. Accordingly, the foreign trade of these companies is carried out by simple export.

To analyze the effect of different internationalization strategies on the firms' economic performance the EBIT (Earnings Before Interests and Taxes) margin was used as the dependent variable and proxy for the firms' economic success. EBIT is a commonly used performance indicator in the literature as it measures the earning power of a company regardless of its financing structure and tax aspects. The EBIT margin can also be used to compare companies of different sizes by taking a relative view of sales (Qian, 1994; Capar and Katobe, 2003; Li, 2007; Heyder *et al.*, 2011).

As control variables they used legal form, firm size (represented by turnover), milk price index as proxy for the price of milk products, gearing ratio and 5 products' variables (fresh milk, dried milk products, fresh milk products, butter and cheese), representing the product portfolio of the firms as control variables. For analyzing the panel data they used a Hausman Taylor estimation with clustered standard errors to take endogeneity, autocorrelation and heterogeneity issues into account (Meyer *et al.*, 2019).

$$Y_{it} = X1_{it}\beta_1 + X2_{it}\beta_2 + Z1_i\gamma_1 + Z2_i\gamma_2 + \alpha_i + \varepsilon_{it} \quad (1)$$

Where Y_{it} is the dependent variable of Firm i at Time t. $X1_{it}$ are time-varying and $Z1_i$ are time-invariant variables that are assumed to be exogenous and not correlated with α_i and ε_{it}. $X2_{it}$ are endogenous time-variant and $Z2_i$ are endogenous time-invariant variables as they correlate with α_i, but not with ε_{it} (Baltagi *et al.*, 2002). Thereby α_i denotes the time-invariant term of the error term, while ε_{it} denotes the remainder component of the error term which is uncorrelated over time (Verbeek, 2004).

The combination of secondary and primary data gives us significantly higher analytical sharpness. On the one hand, we can empathetically demonstrate how different internationalization strategies and other factors affect the economic success of companies by drawing on secondary data in the form of their annual financial statements. By combining the expertise of the experts interviewed in both countries, we can also identify more recent problems, strengths and weaknesses in the dairy industries in both countries and all of this in a much higher level of detail and with aspects that are not available when using secondary data. In sequence we propose ways for the integration of both sides, based on our knowledge of the contexts. More specifically we propose strategies for the internationalization of German processing companies that are willing to invest in the development of the GFM dairy supply chain, in order to enjoy the benefits of this great potential agribusiness partnership. We also derive policy recommendations for the Brazilian government in order to facilitate the entry and actions of such companies in GFM.

4 Results and Implications

4.1 Farm level Brazil – Low Professionalization Level

On farms the **low professionalization levels** are causing several problems. Fifteen interviewees agreed that one of "*the weakest point in the chain is the professionalization of farm management*" (DSBR06) and that this significant problem needs to be fixed.

"**Quality and sanitary aspects** *are issues that must be improved in the supply chain" (DSBR05).* The indices are too *"...variable and difficult for industry standardization" (DSBR08).* Ten interviewees agree that an improvement of these parameters is essential, the implementation of inspection and quality control systems will be especially important if the industry aims to reach international markets. *"In order to export, the country has to develop a program of quality improvements to reach the international standards" (GOVBR02).*

One of the most frequently mentioned problems in the sector is the **establishment and enforcement of contracts** between producers and processors. Most transactions are done on the spot. At the processing level, there are disincentives to use contracts with small producers due to the high transaction costs involved, which are associated with providing inputs, credit, extension services plus product collection and grading, including the time and costs involved in the enforcement of such contracts; all are, once again, major discouragements. Almost all the processors interviewed, stated that they currently work without contracts with producers; some of them had used contracts in the past but not anymore. *"There is also a problem with the seasonality of production, which makes it still harder to sign contracts" (DSBR05)* where *"...production and prices are instable along the year." (GOVBR02).*

The region also faces problems with high costs related to the transport of milk from the farms to the plants. The main reason is poor **infrastructure** related to *"... the bad situation of the roads" (GOVBR01)* in addition to the large distances involved in collecting low volumes from small producers, which *"make the cost of milk procurement higher" (USBR02).*

In GFM, companies are dealing with high transport costs problems in different ways. First through technical assistance in order to raise farmers' total production. Collecting more milk from each farm reduces time taken for collection so optimizing the amount of milk collected per day, per truck and per driver. To facilitate the collection, companies also installed cooling stations in centralized areas, where the milk is brought by farmers or in small trucks belonging to the company. From these cooling stations, larger trucks take the milk to processing plants. Secondly, outsourcing the milk collection is also a strategy being used in GFM. Companies want to avoid the complexity involved in milk collection and focus on their core competencies, letting transport to specialized companies. Transport requires an important mobilization of people in logistics, driving, buying and repairing trucks and inspection controls for the trucks, amongst others. Processing companies also use lobbying with local authorities for improvement of road infrastructure.

4.2 Dairy Industry Brazil – Progress Affected by Low Professionalization

One of the main findings of our research is that **low professionalization of human resources** in production and processing levels in the dairy supply chain may be the principal cause of several other problems. Lack of skilled labor in processing plants and management teams are perceived as huge problems. The **transmission of technology** and best practice for farmers is also affected by low professionalization at the processing level. Companies' managers must be highly qualified themselves in order to provide such assistance. Quality and productivity at the farm level depend upon frequent and effective technical assistance plus diffusion of technologies from processing companies. Current poor quality assistance and its infrequent offering slows modernization in the chain. Interviewees recognized that "*the only way to improve [competitiveness] is by increasing productivity and making farms' production viable* (DSBR09) *through good farming techniques and animal genetics*" (GOVBR05). This would only be possible through the diffusion of technology and techniques to producers. However only a small proportion of farmers are sufficiently specialized in dairy production, in some cases making necessary investment unaffordable. This factor also seems to be hindering the competitiveness in GFM.

Some companies claim to offer a technical assistance, having *"... a department for the promotion of quality, nutrition, silage, hygiene" (DSBR09)* or even a *"a program of technical assistance to reduce the problem of seasonality. They work in the pasture, nutrition, pregnancy rate in the summer to search for stability in the production" (DSBR06).* In total thirteen interviewees stated that the sector offers technical assistance – though it is precarious and lackadaisical – and the most frequently mentioned fields are quality and hygiene, and animal nutrition. In addition, most of the entities, which offer this service, are cooperatives, underlining the importance of these organizations for technology diffusion and farm management.

Processing companies also face high **idle capacity** rates in their plants, which are costly and also generate inefficiencies. Idle capacity affect processing plants that do not have sufficient milk suppliers because they cannot offer better prices to producers, cannot manage the seasonality or simply because of an excess of infrastructure or poor management and planning. Seven people interviewed noted this problem. *"Many industries are still working with idle capacity, which 'weighs' the production system" (GOVBR09).* They mention that *"this idle capacity is very costly" (USBR03)* generating losses and inefficiency. It is also linked to seasonality, disloyalty and control of supply, and poor management planning.

In terms of investments, five interviewees, representing large cooperatives and private companies, argued that there is low level of **investment** in the sector, especially in **marketing and Research, Technology, Development and Innovation (RTDI)**. They say that in general, managers still consider marketing an expense rather than an investment, arguing that *"there is a very poor culture of investment in RTDI and marketing" (GOVBR07)* as a consequence of non-professionalization in the chain. Only six participants mentioned marketing as an important investment. Seven participants affirm that they **invest in RTDI** to improve competitiveness, however *"there is still a huge gap to improve and create more products, companies should also diversify the presentation of products, types and sizes of packages" (GOVBR02).* In terms of differentiation, only two interviewees confirmed that their companies had implemented product differentiation as a strategy; a further four believed that companies had difficulties in differentiating products but should so to increase their profits, especially *"micro and small companies should differentiate products in order to have gains in the niche markets" (GOVBR05).*

4.3 Internationalization Strategies in the German Dairy Industry – Impact on Economic Performance and its Implications

As Table 3 makes clear, companies assigned to individual internationalization strategies differ in important characteristics. Companies with multinational strategies are more profitable than those with international and global strategies. On average, companies are also larger in terms of turnover. However, the average is raised by the Danish and Dutch dairies included.

The number of countries with subsidiaries refers to countries in which the companies maintain subsidiaries and hold at least 50%. With the exception of companies with a multinational strategy, these subsidiaries are only used for distribution and sales by the other companies. In terms of the number of brands, companies with a multinational strategy are also ahead. However, this depends to a large extent on the companies' brand policy. For example, a company that sells its products on all markets under one brand, might offer other brands adapted to the respective countries. The share of foreign sales in total sales (FSI) is also highest among companies with a multinational strategy, followed by those with a global and international strategy (Meyer *et al.*, 2019).

German dairies which follow the multinational strategy are all private ones. The only cooperatives in this group of firms are the Dutch dairy FrieslandCampina and the Danish/Swedish dairy Arla. Both a private and a cooperative dairy are pursuing the global strategy. Of the ten firms following the international strategy 7 are cooperatives (Meyer *et al.*, 2019).

Table 3. Characteristics of groups of firms considered with regard to the internationalization strategy

Strategy/Firms name	EBIT margin Ø2010–2017	Turnover Ø2010–2017 (m. €)	No. of countries with subsidiaries (2017)*	No. of brands (2017)	FSI (2017)
International strategy					
Minimum	0.2%	183.6	1	1	0.0%
Maximum	3.1%	5,089.8	10	11	47.7%
Average	1.4%	896.4	3	5	33.3%
Global strategy					
Minimum	0.8%	879.0	5	15	45.6%
Maximum	4.7%	1,386.9	10	16	56.5%
Average	2.7%	1,133.0	8	16	51.1%
Multinational strategy					
Minimum	3.2%	710.6	8	1	49.7%
Maximum	4.9%	10,756.1	44	46	76.6%
Average	3.9%	4,546.5	22	20	62.0%

* including Germany
Source: Authors' own depiction based on Meyer *et al.* 2019

As the results of Meyer *et al.* (2019b) show the multinational strategy has a highly significant and positive influence on the firms' economic performance (see Table 4). Beside this, the production of cheese and dried milk products also show highly positive influence on the firms' economic performance. It is also noted that the gearing ratio has nearly no influence on the firms' economic success.

The results show that the benefits of locality and differentiation exceed those of globalization and standardization in terms of companies' economic success , according to the EPRG model. According to Meyer *et al.* (2019b) the reasons for this can be seen in the ability of dairies to offer more differentiated fresh milk products, for example yoghurt and fresh cheese, whereas this is not possible for dairies following the international strategy in distant markets due to the limited durability of the products. Beside this, firms following the multinational strategy seem to have advantages in adapting to local demand conditions and building up brands which will lead to higher turnover (Meyer *et al.*, 2019; Harzing 2000). Other studies, such as those of Qian 2002, confirm the positive effect of multinationalism and product diversification on economic success.

Table 4. Influence of analyzed factors on the firms economic performance

EBITmargin	Coef.		Robust Std. Err.	P>z
TVexogenous				
Turnover	0.000		0.001	0.921
Gearing Ratio	0.001	**	0.001	0.048
Dairy-Price-Index	-0.008		0.007	0.268
Fresh milk	0.014		0.374	0.971
Fresh milk products	0.528		0.594	0.374
TIexogenous				
Legal form	-0.142		0.825	0.864
Cheese	2.881	***	0.823	0.000
Dried milk products	1.528	***	0.456	0.001
Butter	-2.157	***	0.747	0.004
TIendogenous				
Intern. Strategy	3.102	***	0.539	0.000
Constant	0.149		1.449	0.918

*p<0.1 **p<0.05 ***p<0.01

Wald chi2(10) = 507.21; Prob > chi2 = 0.0000; sigma_u = 0.5739; sigma_e = 1.2511; rho = 0.1738

Std. Err. Adjusted for 18 clusters in ID; TV = time varying variables; TI = time invariant variables

Source: Authors' own depiction based on Meyer *et al.*, 2019

4.4 Implications for the German Dairy Industry

Our analysis shows that the German dairy industry faces various challenges. These result in particular from increasing and changing requirements of milk production. After the milk quota came to an end, other factors, particularly stricter regulations under the new Fertilizer Ordinance, are likely to become the limiting factors (Meyer and Theuvsen, 2017). In 2020 the restrictions will be further tightened (Top agrar, 2020). This applies in particular to intensive regions of dairy cattle farming and thus to the growth of milk production in the most important regions. Beside this there are current trends such as increasing quality orientation, sustainability, transparency and the increasing demand for organic food that offer companies the opportunity to position themselves here with appropriate products, the whole process is taking place against the background of a saturated domestic market (Bratschi and Feldmann, 2005; Brümmer *et al.*, 2018; Mehlhose *et al.*, 2019).

As mostly any increase in ecological production, increasing requirements with regard to animal welfare, etc. will make production in Germany more expensive, the dairies have to relativize the additional costs by means of differentiated products and higher revenues. On the international milk markets it will be very difficult or even impossible to achieve the necessary extra price for these products, since both within the European Union and outside it, the most popular product characteristic is a competitive price (BVE, 2017). This applies in particular to less differentiated export products such as butter and milk powder but also to standard cheeses. Other studies also show that companies that pursue an export strategy of little-differentiated goods have significantly poorer value added and economic performance (Heyder *et al.*, 2011; Jürgens *et al.*, 2015). Beside this, market shares of Germany are decreasing over all product groups except for butter in the last years. Even increasing quota amounts could not change this development (Trademap, 2019). In addition to the aspects mentioned above, this situation becomes even clearer if one looks at the projections on the development of milk production and demand. They predict average production growth of 1.7% p.a. for the next decade, with more than half of this in India and Pakistan. Despite the shorter shelf life, most milk will be consumed in the form of unprocessed or poorly processed fresh milk products. Average production growth of 1.4% p.a. is forecast for Latin America, with the focus on the southern countries and Brazil. The region remains a net importer of milk and dairy products (OECD and FAO, 2019).

Against this background, we believe that GFM offers companies in the German dairy industry good prospects for future growth and market development. The above analysis shows that, despite the opportunities, commitment is not without its problems and risks. Whether and under what conditions it can nevertheless become a win-win situation for both sides is discussed in the following chapter.

5 The Integration of Supply Chains to a Win-Win Manner

5.1 Competition, Market Entry and Strategy Opportunities

GFM is a very dynamic and fast growing region in dairy production. Large companies have been installing plants in the region[3] and competing for the procurement of milk, provoking controversy among interviewees. Some believe that these companies will develop the dairy sector by stimulating improvement in techniques to achieve greater competitiveness. Others think that these companies harm the smallest ones and bring

[3]Nestlé installed two plants in 2008 and 2010. Lactalis arrived in 2014 and is already the largest group in GFM. Other large companies include Tirol, Italac, Piracanjuba, etc. The largest cooperatives are CCGL installed in 2008 and Aurora, which started processing milk in 2004.

about negative consequences by establishing a monopsonistic position in milk procurement from some areas and therefore have an anticompetitive effect. Some interviewers affirm that *"the entry and expansion of* ***large companies*** *increases the competitiveness of the sector" (USBR02)* and it *"brings improvements in competitiveness, boosting production and innovation" (DSBR05)* so that the sector is *"... becoming competitive, professional instead of familiar and that raises the prices [for producers]" (DSBR02).*

Furthermore, as profit margins decline, increasing concentration is inevitable in order to spread fixed costs and remain competitive (Sutton, 2003). Porter (1990) considers that rivalry generates pressure on competitors and stimulates sustainable and continuous growth to maintain competitive advantages. The process of concentration and internationalization is inevitable in the modernization of a supply chain but these processes should happen in a fair way, especially for small producers, who are the most affected. This is often limited in monopolistic situations and particularly in the case of companies' opportunistic actions. In this case the consequences of monopolies would be lower returns to farmers, increase risk in farming activities and cut-off more farmers and small companies, especially cooperatives. In this regard mergers and acquisitions of small companies would be an important strategy to realize gains in scale and bargaining power. However, professionalization of the management of such companies is also fundamental. Concerning the cooperatives, interviewees mentioned that *"... there is a tendency for any merger between cooperatives to compete in scale" (DSBR01)* and to improve their cost structure and reduce idleness in processing plants through more efficient planning. Despite interviewees' awareness, we did not see much evidence of these mergers in GFM; they should occur faster and involve more cooperatives. Various successful examples of mergers and acquisitions all over the world reinforce this strategy, such as Fonterra, DFA, FrieslandCampina, Arla and others.

Particularly in view of existing idle capacities, entry into the market through takeovers or mergers seems to be the preferred method rather than new construction. This was also the way taken by Lactalis and Ehrmann[4], for example, which chose this path when they entered the market.

In terms of business strategies, GFM and Brazilian's conditions allow companies to implement both **a cost leadership** and a **differentiation strategy**. On the one side the

[4] Ehrmann entered the Brazilian market in 2018 through a joint venture with the Brazilian dairy Trevo Lacteos S.A.

internal market has a high demand potential and so far, a low offer of products differentiated. On the other hand, low prices for inputs, labor, land and good climatic conditions offer an attractive scenario for cost leadership strategies and scale gains with the focus on exports. There are companies in Germany that implement these strategies. The cost leadership strategy is usually implemented by large cooperative dairies from the north, while the private dairies from the south pursue a differentiation strategy. "*There is a big difference in Bavaria, because compared to Lower Saxony, there are special types of cheese there. Bavaria offers more specific products than northern Germany with standard cheeses and skimmed milk powder that anyone can produce. They have higher costs but in the end they also have better products. When you think of the German milk market and the well-known brands, they are particularly Bavarian*" (GOVGE01). They offer products with unique attributes such as well-known brands, superior product qualities, innovation, reliability, environmental or regional appeal. The most important types are organic milk, non-GMO milk, hay milk, grazing pasture milk, mountain milk, the *Geprüfte Qualität Bayern* label and the animal welfare label. The value added is correspondingly high at 1.09 €/kg milk. However, the export focus is lower. Of the € 9.07 billion turnover generated in 2016, 27% was generated from exports (11% of German exports). There are a total of 82 milk-processing locations in Bavaria, which leads to competition between dairies for milk. Thereby 50% of milk is processed in cooperatives and 50% in private dairies.

In Lower Saxony 78% of the milk collected is processed in cooperative dairies and 22% in private dairies. Lower Saxony dairy companies primarily pursue a cost leadership strategy and try to use favorable production conditions to sell their products on the world market. This is demonstrated on the one hand by the significantly higher export share of their own production compared to Bavaria; of € 4.05 billion turnover in 2016, 40% was attributed to exports (6.8% of German exports). On the other hand, the products are not very differentiated, which also underlines the significantly lower added value of 0.64 €/kg milk compared to Bavaria. In Lower Saxony, the largest German dairy cooperative, Deutsche Milchkontor (DMK), dominates, followed by the second largest dairy cooperative, Ammerland. There are only 14 milk processing sites in Lower Saxony, which significantly reduces the competition in milk compared to Bavaria "*The dynamics and relations with the producers in Lower Saxony are quite different from those in Bavaria. In general, the cooperative does not have the same concern for farmers: There are far fewer competitors, there are some private companies and cooperatives, but DMK is the dominant one*" (DSGE02) "*Many dairy companies have brands with high innovation potential and we find that in Germany, in southern Germany, Bavaria is still the region with the highest milk prices, because of these high value-added products, but farmers have many opportunities to change dairies too* (supplier competition) (DSGE03)".

Due to the fact that companies will strive for high value creation and that production costs in Brazilian dairy production are still comparatively high due to the above-mentioned problems, companies entering the market should implement a differentiation strategy focusing on the internal market potential. Studies show, that this is this has to be favored when it comes to added value. In addition, the companies could advertise with their German origin, which enjoys a high international reputation and is linked to attributes such as high quality and food safety. (IFCN, 2013; Jürgens, *et al.*, 2015; BVE, 2017; Brokelmann and Gäde, 2017).

The concentration in the supermarket/retailing sector should support the differentiation strategy. Though it has consolidated rapidly in Brazil over the past years, mainly via mergers and acquisitions, in 1994 the 10 largest supermarket groups had 24.3% of market share, while in 2018 the 10 largest have 42.2%, it is far less concentrated then in Germany (ABRAS, 2018; Farina, 2002). While eight large retailers in Germany accounted for 70% of market share in 1999, the four largest food retailers achieved a market share of 85% in 2011. Accordingly, manufacturers' negotiating position vis-à-vis food retailers is comparatively unfavorable in Germany (Schlippenbach and Pavel, 2011). In view of lower concentration levels in the Brazilian retail sector, this should have a positive impact on the dairies' bargaining position which may have a positive impact on conditions, such as prices, placing and listing of products.

5.2 Coordination and Organizational Forms

In addition to the question of how companies can structure a market entry, the question also arises of which companies are even considered for such a step. The German companies that pursue a multinational strategy are currently all private dairies. While cooperative dairies like Arla or FrieslandCampina have been implementing the multinational strategy for some time now, there is not a single German cooperative dairy doing the same (Meyer *et al.*, 2019). The biggest obstacle in this respect is probably the attitude to risk amongst cooperative members and managers. This is most likely to be the case, when in particular, the dairies' disbursement prices are comparatively low. Beside this, free-riding, investment, transaction costs, controls and human resource problems can also be reasons (Theuvsen and Ebneth, 2005). Indeed, there are differences in **governance structures** concerning cooperatives. The two largest German cooperatives, Bayerische Milchindustrie eG and DMK, have created a governance system, which means that farmers, i.e. the actual owners, no longer have any influence on the operative business. To manage the cooperative, DMK has set up a limited liability company, DMK GmbH, which is made up of external experts such as the Managing Director and the Chief Financial Officer appointed by the cooperative's Supervisory Board. The Supervisory Board therefore controls not only the rights but also the central elements of the cooperative. As a result, DMK's management structure reduces the decision-making authority of those members and transfers it to the Board of Management. Governance structures that virtually deprive cooperative members of their operational power could like in the case of the DMK, enable some cooperative dairies to raise the necessary capital for such a step by not putting it into the remuneration of the milk delivered. "*At DMK there is a cooperative and a company at the head of the company. The cooperative collects only the milk and receives money from the limited liability company, which it distributes to the farmers. Thus the farmers, who are at the head of this cooperative, cannot influence the GmbH*" (GOVGE04).

Therefore, theoretically both cooperatives and private German dairies are in position to establish themselves in GFM to develop their internationalization strategies. The Ehrmann dairy, for example, has already taken this step. For the time being, an entry into the Brazilian market, at least in the short term, is likely to be reserved for some German private dairies due to governance structures and economic efficiency due to the high factor demands in terms of equity endowment and financial power. Here the cooperatives have a clear disadvantage in relation to the private dairies (Brokelmann and Gäde, 2017). In GFM we observe a balance between the share of milk collected by cooperatives (46.5%) and private companies (53.5%) (Beber *et al.*, 2018). With regard to cooperatives, the traditional cooperative system is predominant, since the Brazilian Federal Law 5764 of 1971 is not flexible enough to allow other organizational structures in the cooperative system.

In Germany we observe good coordination among farmers and processing companies in the dairy sector, with the balance of power ensured by governance mechanisms in the supply chain. On one side the cooperatives have greater flexibility and have developed modern management structures, so that the price of milk is negotiated mainly within the cooperatives owned by the producers. To a certain extent such structures can be applied in Brazilian cooperatives. On the other side, we observe strong coordination amongst farmers who negotiate their milk prices with private companies via Producer Organizations – POs (MeG – Milcherzeugergemeinschaft), mainly in Bavaria. Such POs guarantee a well-functioning system in imperfect markets. These allow for the negotiation of entire volumes of a group of producers with the dairies. In turn that ensures greater bargaining power with the producers and consequently higher prices and better sharing of value added. Whether this could prove an interesting solution for Brazil will depend heavily on the presence of several competitive dairies, each competing for milk as is the situation in Bavaria (Jürgens *et al.*, 2015). Furthermore POs already exist in GFM; the ASCOOPER and SISCLAF are examples (Beber *et al.*, 2018).

5.3 Professionalization of the Human Resources – Quality of Agricultural Training, Technical Assistance and Advisory Services

The effective development of a supply chain is achieved by the work of qualified professionals. Therefore, all the aforementioned aspects would only be possible when highly skilled professionals reflect about the best ways for their applications to be context-adapted and then take action how to actually implement them. In Germany, the dairy supply chain, and agriculture in general, developed and received high levels of support. Investments in R&D and the system of vulgarization are very functional and effective. Professionals have to be qualified in order to remain active and they are indeed well qualified. The use of public and private technical and economic advice seems to be quite sought after by milk producers. "*There are a handful of consultants, people who are important to us* (FAGE04)." "... *we are all well-educated and you need a consultant because he has accurate data on other farms* (FAGE08)".

The German Agricultural Information and Knowledge System (AKIS) is made up of a wide variety of organizations and institutions with a long tradition and well-established roles. All categories of organization (public administration, public and private research and education, private sector, farmers' organizations and non-governmental organizations) are represented. According to the results of the 2013 agricultural census, 68% of all agricultural managers had received agricultural vocational training. The remaining

32% only had practical agricultural experience. In large enterprises, almost all farmers have received agricultural vocational training. Of the farm managers who completed their agricultural vocational training, 11% hold a university degree. Figures in Brazil show a different situation at the moment, with only 1.35% of farm managers having vocational training (agriculture technician) and 4.16% holding a university degree. Agricultural vocational training is not widely developed across the country.

In this regard, GFM dairy professionals should receive constant training and advice in the best practices of dairy farming and processing. First, applied Research and Development (R&D) should be further enhanced jointly by companies, universities, associations, extensionists and other R&D institutes for the identification and development of best practice, breeds, production systems, feed, etc. In sequence, best practice has to be transferred to dairy professionals so ensuring their application. In GFM, institutes such as EPAGRI, EMBRAPA, EMATER, universities and others have already developed important work in this regard and contributed to the improvement of dairy production parameters over the past few years.

In partnership and supported by these institutes, processing companies are developing programs of agricultural training and technical assistance for farmers in GFM and investing in R&D to generate better technology for the sector. Some companies "*release new products every year, have a department of innovation and R&D*" (DSBR05) or even run "*[on the cooperative] an experimental center to develop technology for pasture-based milk production. Also, some have an experimental dairy farm*" (DSBR01). Those practices must be further supported by processing companies in order to take advantage of the full potential that the GFM resources have to offer.

Despite the progress made in this area, companies planning to enter the Brazilian market should at least start by training their employees and, if necessary, the farmers themselves to be made aware of the issue. For example, firms could try to bind employees to the company by offering them vocational training in Germany and, in return, committing them to work for the company for several years.

With regard to training farmers, there have been exchange programs for some time now, such as the Deula[5] exchange program, in which Brazilian farmers learn on German

[5] The educational institutions affiliated in the Federal Association DEULA e. V. belong to the most important regionally and supra-regionally active agricultural technical educational institutions in Germany.

farms. These could be further promoted and supported by the dairy companies in order to improve both education standards and performance on their dairy farms. In contrast to other regions, which are also forecast to experience strong growth in milk production and demand, GFM offers German dairies in particular the advantage of cultural proximity. Despite the declining trend, many people still speak German. This aspect and the cultural proximity should have a positive effect on the training and exchange of employees and farmers.

5.4 Milk Quality and Sustainability Parameters

Regardless of the dairy's organizational form, any market entry would require simultaneous support and advice from processing companies to their suppliers, especially with regard to production technology and professionalization.

One of the first measures on which processing companies have to work, is to enhance the quality of the milk produced by the farmers. Besides controlling for several problems during milk processing, better quality ensures better products and improved market access. If quality standards cannot be improved or adhered to, there is a risk that companies will jeopardize their good reputation, or that of their brands, and cause themselves considerable damage. Programs for farm training and technical assistance would be essential together with a program for the payment for quality and solids. Payments for quality are incipient in the region but are growing with the instalment of new companies. ... *"there is a tendency for payments per quality and solids because that's only what interests in milk production" (USBR03).*

In order to attract farmers, interest in investing in technical progress and a good **contracting system** should be enhanced. Its benefits then need to be communicated to farmers, with the aim of ensuring their loyalty and compliance to the quality requirements. It gives farmers the opportunity to improve and can reduce price variations, increase incomes for poor farmers and promote rural development (Key and Runsten, 1999; Andri and Shiratake, 2005; Alemu and Adesina, 2015; Koester and Cramon-Taubadel 2019). However, it must also be associated with an efficient program of technical assistance for quality improvements. Moreover, improvements in the judicial system (easier access and reduction in bureaucracy) to enforce and make that enforcement of contracts cheaper, may also help to solve the problem. This point must be promoted

by local authorities in order to promote the development of rural areas with higher value added and the creation of jobs.

Looking at the German case, we see a great deal of experience in quality management in the supply chain as a whole, on which processing companies can be based to promote necessary quality enhancement. The main mechanism in place in the country for quality management is the QM-Milch Association (Qualitäts Management Milch). It is a national mechanism founded by the farmers' association, the Raiffeisen association and the dairy industry association, which is also supported by the entire dairy industry and is responsible for setting standards between milk producers and dairies. QM Milch sets strict quality standards, which are verifiable for milk production and apply uniformly throughout Germany. They specify requirements that go beyond legal requirements and those of good professional practice. With its standards, QM Milch ensures that not only the quality of the product is guaranteed, the entire production process is transparent and traceable. Other sustainability parameters such as environment, animal welfare and socioeconomical characteristics are under development to be included in a broader QM Milch assessment. The majority of German farms are part of this scheme, "*over 90% in some regions, must meet these standards of quality milk. It's a neutral certification, neutral auditors go to the farms. If you do not meet the QM standard for milk, especially in northern Germany, you will not find a dairy that collects your milk* (GOVGE01)".

Since the creation of such a standard is a lengthy process[6], dairies entering the Brazilian market would initially be left to their own devices to set standards and control them. However German companies have experience in such a company-owned initiative such as the "Milkmaster" program of the DMK for example. This important aspect should also be promoted nationwide in cooperation with Brazilian associations and authorities in order to increase the transparency of the value chain throughout the country. In Brazil, public regulatory standards set the requirements of milk quality. The Normative Instructions 51, 62, 76 and 77 of the Federal Government regulate such standards in Brazil.

[6] The groundwork for the QM Milch was laid in a working group in 2002, while it was founded officially in 2011.

5.5 Farm, Herd and Feed Management

An essential point for successful engagement in Brazil will be the support of agricultural enterprises with regard to production. Farm, herd and feed management parameters must be controlled and also be viable to ensure competitiveness for farmers and the whole sector. This process has to be led by processing companies but coordinated with local advisory and consultancy institutes. In addition to the advantages gained for the agricultural enterprises and the securing of raw materials, it would also improve the quality of the milk and thus support the above-mentioned aspects with regard to quality assurance.

In 2006 the average productivity was 2,414 kg/cow/year in GFM. Extensive work to improve such parameters have been conducted over the past several years to develop the dairy supply chain. Processing companies, R&D and extension services institutes supported by the local governments helped farmers to achieve productivity of 3,855 kg/cow/year in 2017 (IBGE, 2018). These numbers are approaching those of New Zealand (4,500 kg/cow/year), which has similar pasture based systems in dairy production, transforming the great efforts of dairy professionals in GFM. However, there is still a lot of work to be done at farm level to improve the parameters of dairy production, which are achieved through better management and technology adoption. In this regard, German dairy professionals are better prepared and can contribute in different ways. We identified several good aspects of farm, herd and feed management in Germany, which could possibly be extended to GFM.

About 74% of dairy farmers in Bavaria and 73% in Lower Saxony have at least some training in agriculture. In the interviews we noted that there is impressive **management capacity** and **economic monitoring** among the producers and that the management of the farm and its techno-economic indicators are generally well monitored by the producers. Despite notable regional differences, cost management remains a strategy adopted at the national level. Cost leadership strategies are very important. In this respect, actions can involve the ease of operationalization of certain activities, gains in technical and allocative efficiency, reduction of labor costs, choice of more suitable breeds and outsourcing activities, among others. *"Cost reduction is the most important task to maintain or improve the competitiveness of a dairy farm. Get the best with the least effort possible (FAGE06)." "To stay competitive, it is essential to reduce unit* *costs in certain areas. We want to achieve high performance in dairy farming with lean production and with little effort. We have been very successful in realizing this principle in recent years. Lean production means comparing the prices of mineral foods from different producers, etc. These are just small set screws, but it helps to survive a crisis* (FAGE07)."

"*At the milking robot, an incredible amount of data is measured, for example the electrical conductivity of the milk, which is then listed and listed in the cows. We check if the values are correct and examined if the cows have all recovered their food concentrate. When the milk is collected and the parameters (number of urea and cells) are measured, whenever something goes wrong, we immediately check if there are any outstanding animals in the herd. ... in some cases, we have already compared with the adviser of my Beratungsring, then you have the list of the 25 best producers. And we usually exchange, for example, during information days and of course on the Internet;* (FAGE03)."

In Germany, **digitalization** and **technical progress** of farming activities and processes are gaining significant attention amongst dairy farmers, especially the new generation. The patrimonial strategy for those who continue in the activity, and especially for those who have an expectation of inheriting the farm, pursues a strategy of intensive capitalization, expansion and digitization. Investments are generally progressive and constant. "*You see that farms are becoming more capital intensive, they are investing more in machines, and so on. This is a normal trend that you see everywhere* (POGE1)." "*For now, it's important to modernize the business with a new stable concept* (FAGE10)." "*So they can invest in milking robots and some farmers invest in cutting-edge technology for milking, automatic feeding, etc., then harvesting the grass, etc.* (DSGE03)".

The work in breeding is also more and more automated. Automated systems are already widely used in the barn. These include milking robots, sensor-based call flows or automatic feeding systems. Robots are also already used to provide basic feed, clean running surfaces and convert grazing fences. Milking via the automatic milking system (AMS, also called a milking robot) has developed rapidly. In 2016, around 7,800 milking robots were used on 5,500 farms in Germany. Milking robots have been part of the state of the art for years and two out of three dairy producers now opt for an automatic milking system for a new purchase. "*Something had to happen here and after consulting my son, we moved to milking robots. Because he (the son) wanted to work with robots during his acquisition because he could save a workforce and you have incredible flexibility in terms of working hours* (FAGE01)." "*... yes, not only because of less work, it may be less, but it's easier work, more attractive work. Young people like to work with laptops, etc., and not in the barn. That's what I see at the moment, those farms that have invested in automation, it's easier for them to hire qualified staff* (DSGE01)".

In this regard the local government in GFM must foster the technology adoption through the availability of credit lines for financing the purchase of such equipment (technology improvement), and processing companies should provide advice on how to best use these new technologies (efficiency improvement). Beside this, the mentioned exchange programs for farmers could support the progress of digitalization and technical progress, as young farmers get in touch with these technologies.

Other strategies for cost reduction are **collective actions** which are also put in place in Germany, such as 'machine networks'. "Maschinenring" are associations that allow producers to rent the necessary machines, without the need to buy them, thus reducing investment costs, maintenance and depreciation, amongst other cost factors. A producer can also offer his services through a "Maschinenring". "*Maschinenring is a common practice here, and all over Germany, when I read some of the agricultural newspapers, "Maschinenring" is spread everywhere. You can reduce your investments and your fixed costs* (DSGE03). By advising several dairy farms in a region, the sharing of machinery could be supported or encouraged by dairies to give farmers faster access to technical progress.

Work on necessary technical assistance should be developed by the processing companies and local institutes together with farmers, so best practice for **herd and feed management** common in Germany could be widespread in Brazil. The management of cows with technical and performance indicators is recurrent. "*I've been rigorously planning how we manage cows for calving because our calf mortality rate is a bit high. Daily and weekly controls, what cows eat, so we have strict management and we know what to do, so everyone knows exactly what to do to improve control of everything* (FAGE03)."

"*In Germany, the percentage of insemination, I think it's about 90% ... Insemination is usually done by professionals from outside. 80% of inseminations are done by our tech-*

nicians, 20% by farmers. And we offer every year 2 educational weeks for farmers for this (USGE04)".

Registration and monitoring of livestock by associations is also common, "*I would say that 80% of the cattle in our area are registered cattle". "I think what's really important is that we started 15 years ago to provide our farmers with reproductive services, to carry out pregnancy checks. ... farmers learned that it was necessary to put in place a strategic breeding control, to come every 2 weeks, 4 weeks to control cows after calving, to control gestation, look for problems in cows and we started very early as a center of Artificial Insemination to offer these services* (USGE04)".

For feeding, different strategies are employed, batch separation according to nutritional requirements, but also unique batches, individualization by cow with robots, TMR (total mixed ration), autonomy of production of food in the farm, among others. But above all, nutritional quality emerged as a factor considered very important.

"*The quality of the feed has to be right. If you make bad feed in the summer, you will notice it immediately in winter in the milk. Either in quality, in terms of number of cells or performance. We do not balance much with concentrates, etc. This means that complete production depends on the quality of the base feed* (FAGE02)."

Feed costs are the most important among specific livestock costs. They account for about 60% of the total specific costs in the two study regions. Production autonomy is also highly sought after by producers. "*In our individual case, we produce more than we need and we sell the rest* (FAGE02)". In view of the great importance of feed costs, dairies should offer farmers good advice here. The production costs in comparison with other countries show that Brazilian agriculture still has a lot of potential for improvement (IFCN, 2013).

6 Possibilities and Necessity of Political Support

Dairy companies that intend to seize such opportunities in GFM and are willing to contribute to the development of the dairy supply chain must receive support from the local governments in order to facilitate the process. The respective government must define and work on a projects to develop the dairy sector, as was the case for other agricultural sectors in Brazil that are much more competitive, such as soybean, meat and sugarcane for example. Such incentives should come as direct support programs via the development of R&D, technical assistance and farming training through public institutes, also as tax incentives and availability of credit lines to companies as financial incentives for the implementation of these actions. Processing companies could act as promoters of the rural credit systems, supporting farmers to get access to the available credit program lines from the government. In the case the dairies pursue a contract system, farmers could use these contracts to get better access and conditions for loans, by showing these contracts and therefore their market engagement to the financial institutions. The government must also take action in strengthening institutions that will guarantee the enforcement of laws, rules and contracts as well as in infrastructure development for transport and electricity.

Ten interviewees, including institutions, cooperatives and private companies, conclude that the problem of high transport costs could be improved by developing the supply chain. "*Increasing volumes per farm and maintaining good routes to access the farms can solve those problems" (DSBR03).* This issue involves the provision and maintenance of the public work service by the local/state governments, over which the producer has low or no direct control. A reliable and stable provision of electricity on the farms is also a fundamental infrastructural ambition since dairy farms have to keep the milk cooled for at least two days before collection (Escher, 2011). In terms of governmental investment programs for example, the state of *"Santa Catarina has invested a lot in technology and genetics. It is also the only state free of foot-and-mouth disease without vaccination" (GOVBR07).* The capital for such investment comes mainly from governmental development banks, or in some cases, the capital is a mixture of both credits and own capital.

These actions must be strengthened and coordinated with the processing industry for the sake of the whole supply chain.

7 Conclusions

In future, the increase in milk production will take place mainly outside Germany and the European Union. In order to maintain or increase market shares worldwide and achieve higher added value, German dairies will be dependent on processing milk outside Germany as well. In addition, competitiveness vis-à-vis competitors on international markets will deteriorate as a result of increasing environmental and animal welfare

requirements for domestic production while the added value of these exports is comparably low anyway.

GFM offers German dairy companies considerable potential. This applies both in terms of available resources and in terms of demand. German dairies would in turn bring know-how and added value to the south of Brazil.

On the other hand, however, there are also challenges that should not be underestimated in terms of raw material quality, the level of training of potential employees, etc. Even if there is progress in Brazil in this respect, companies would be largely on their own if they entered that market, at least in the beginning, when they would have to promote such initiatives. In contrast to Germany, where consulting and training of the supplying farmers do not fall within the area of dairy companies, the situation in Brazil would be completely different and the companies would have to cover this area as well.

Some private German dairies have already gained experience in setting up and producing from dairies abroad. In the short term due to governance structures, this form of internationalization is likely to be reserved for private dairies but cooperatives could quickly follow, if there were intentions for market expansion.

Not only the dairy industry but also the government and the communities should be interested in the development of GFM's natural potential with regard to milk production. Successful development of the milk value chain would increase value creation in urban regions and create attractive jobs. Ultimately, it is also in the best interests of local government to support those companies and farmers who are interested.

Other areas of Brazilian agriculture, such as soya and meat production, have shown that Brazil can be and is extremely competitive worldwide. From the natural resources available, there is nothing to prevent the milk value chain from following an equally successful path. Against this, other resources still have to be developed or further promoted in order to follow this path successfully. There is no question that this development would be challenging but German dairies have the necessary know-how to drive this development forward. In the end there could be a win-win situation for both the German companies as well as for Brazilian farmers and rural communities.

References

ABRAS (2018). Economia e Pesquisa » Ranking Abras » As 500 maiores – ABRAS. ABRAS – Associação Brasileira de Supermercados, available at: http://www.abrasnet.com.br/economia-e-pesquisa/ranking-abras/as-500-maiores/ (accessed 23 September 2019).

Alemu, A.E., Adesina, J. (2015). Effects of co-operatives and contracts on rural income and production in the dairy supply chains: Evidence from Northern Ethiopia. *African Journal of Agriculture and Resource Economics* 10: 312–327.

Andri, K.B., Shiratake, Y. (2005). Empirical study of contract farming system conducted by dairy cooperatives in east java, Indonesia. *Review of Agricultural Economics* 55(2): 73–84.

Anschau, C.T. (2011). Redes Cooperativas da Bovinocultura de Leite e o Desenvolvimento do Oeste Catarinense. Dissertation, Universidade Comunitária da Regiao de Chapecó.

Baltagi, B.H., Bresson, G., Pirotte, A. (2002). Fixed effects, random effects or Hausman-Taylor? A pretest estimator. *Economics Letters* 79: 361–369.

Beber, C.L., Carpio, A.F.R., Almadani, M.I., Theuvsen, L. (2019). Dairy supply chain in Southern Brazil: barriers to competitiveness. *International Food and Agribusiness Management Review* 22(5): 651–673.

Beber, C.L., Theuvsen, L., Otter, V. (2018). Organizational structures and the evolution of dairy cooperatives in Southern Brazil: A life cycle analysis. *Journal of Co-operatives Organization and Management* 6(2): 64–77.

BLE (2019). Milchwirtschaft auf einen Blick nach Kalenderjahren. Federal Office for Agriculture and Food, Bonn.

BMEL (2018). Das Ende der EU-Milchquote – Auswirkungen auf den Milchmarkt. Federal Ministry of Food and Agriculture, Berlin.

Bratschi, T., Feldmann, L. (2005). *Stomach Competence – Wachsen in gesättigten Märkten: Trends, Strategien und Konzepte für Lebensmitteleinzelhandel, Food-Hersteller und Systemgastronomen.* Deutscher Fachverlag GmbH, Frankfurt am Main.

Brokelmann, V., Gäde, S. (2017). Branchenstudie Molkereiwirtschaft 2017. HSH Nordbank AG.

Brümmer, B., Spiller, A., Mehlhose, C. (2018). Der Markt für Milch- und Milcherzeugnisse im Jahr 2017. *German Journal of Agricultural Economics* 67: 119–137.

Brunetti, A., Kisunko, G.B., Weder, A., Weder, B. (1999). Institutional Obstacles to Doing Business: Region-by-Region Results from a Worldwide Survey of the Private Sector. The World Bank, Washington D.C.

BVE (2017). Fakt ist: Lebensmittelexport. Fakt ist No. 6, Fedration of German Food and Drink Industries, Berlin.

Capar, N., Kotabe, M. (2003). The Relationship between International Diversification and Performance in Service Firms. *Journal of International Business Studies* 34(4): 345–355.

Carvalho, G.R., da Rocha, D.T., Gomes, I.R. (2018). O Mercado de Leite em 2017. Circular Técnica No. 118, Embrapa Gado de Leite, Juiz de Fora, MG: 1–28.

Cook, M.L., Chaddad, F.R. (2000). Agroindustrialization of the global agrifood economy: bridging development economics and agribusiness research. *Agricultural Economics* 23: 207–218.

Damke, C. (2008). Políticas lingüísticas e a conservação da língua alemã no Brasil- n° 40 Espécule. State University of Western Paraná, available at: https://webs.ucm.es/info/especulo/numero40/polingbr.html (accessed 29 May 2019).

Destatis (2019a). Employees and turnover in the German manufacturing industry from 2008 to 2016. Genesis-Online database. Federal Office for Statistics, Wiesbaden, available at: https://www-genesis.destatis.de/genesis/online (accessed 5 July 2019).

Destatis (2019b). Exports of German dairy and dairy products from 2008 to 2016. Genesis-Online database. Federal Office for Statistics, Wiesbaden, available at: https://www-genesis.destatis.de/genesis/online (accessed 5 July 2019).

Destatis (2019c). Landwirtschaftlich genutzte Fläche: Über ein Viertel ist Dauergrünland. Federal Office for Statistics, Wiesbaden, available at: https://www.destatis.de/DE/Themen/Branchen-Unternehmen/Landwirtschaft-Forstwirtschaft-Fischerei/Feldfruechte-Gruenland/aktuell-gruenland2.html (Accessed 20 August 2019).

Escher, F. (2011). Os Assaltos do Moinho Satânico nos Campos e os Contramovimentos da Agricultura Familiar. Master thesis in Rural Development, Universidade Federal do Rio Grande do Sul, Porto Alegre, RS.

FAOSTAT (2019a). Production of cow's milk. FAOSTAT. Food and Agriculture Organization of the United Nations, available at: http://www.fao.org/faostat/en/#data/QL (accessed 20 March 2019)

FAOSTAT (2019b). Productivity of countries' cows milk. FAOSTAT, Food and Agriculture Organization of the United Nations available at: http://www.fao.org/faostat/en/#data/QL (accessed 18 March 2019).

Farina, E.M.M.Q., Nunes, R. (2002). A Evalucao do Sistema Agroalimentar e a Reducao de Precos para o Consumidor: O Efeito de Atuacao dos Grandes Compradores, Programa De Estudos Dos Negocios Do Sistema Agroindustrial PENSA. PENSA/FIA/FEA/USP, Sao Paolo.

Farina, E.M.M.Q. (2002). Consolidation, multinationalisation, and competition in Brazil: impacts on horticulture and dairy products systems. *Development Policy Review* 20: 441–457.

Farina, E.M.M.Q., Viegas, C.D.S. (2003). Foreign direct investment and the brazilian food industry in the 90s. *International Food and Agribusiness Management Review* 5(2). 1–16.

Friedrich, C. (2010). Milchverarbeitung und –vermarktung in Deutschland – eine deskriptive Analyze der Wertschöpfungskette. Thünen Working Papers, Institute of Farm Economics, Thünen-Institute, Braunschweig.

Gonçalves de Assis, A., Stock, L.A., de Campos, O.F., Gomes, A.T., Zoccal, R., Silva, M.R. (2005). Sistemas de Produçãod de Leite no Brasil. Circular Técnica No. 85. Embrapa Gado de Leite, Juiz de Fora, MG: 1–6.

Görmann, H., Kreins, P., Zabel, A. (2006). Wohin wandert die Milchproduktion in Deutschland? Agricultural Research Völkenrode.

Goodman, L.A. (1961). Snowball Sampling. *The Annals of Mathematical Statistics* 32: 148–170.

Harzing, A.W. (2000). An empirical analysis and extension of the Bartlett and Ghoshal typology of multinational companies. *Journal of International Business Studies* 31(1): 101–120.

Helfand, S., Moreira, A.R., Bresnyan Jr, E.W. (2015). Agricultural Productivity and Family Farms in Brazil: Creating Oppertunities and Closing Gaps. The World Bank, Washington D.C.

Heyder, M., Makus, C., Theuvsen, L. (2011). Internationalization and Firm Performance in Agribusiness: Empirical Evidence from European Cooperatives. *Journal of Food System Dynamics* 2(1): 77–93.

Hörl, M., Hess, S. (2017). The Export Competitiveness of the European Dairy Industry. Contribution at the XV Congress of the European Association of Agricultural Economists, 28 August – 1 September, Parma, Italy.

Holtbrügge, D., Welge, M.K. (2015). *International management theories, functions and case studies*. Vol. 6, Schaefer-Poeschel, Stuttgart.

IBGE (2006). Censo Agropecuário 2006. Instituto Brasileiro de Geografia e Estatística, available at: http://www.ibge.gov.br/home/ (accessed 26 October 2015).

IBGE (2018). Producao de leite e vacas leiteiras em producao. Instituto Brasileiro de Geografia e Estatística, available at: http://www.ibge.gov.br/home/ (accessed 10 August 2019).

IFCN (2013). Dairy Report 2013. International Farm Comparison Network, Kiel.

Jürgens, K., Fink-Keßler, A., Poppinga, O., Wohlgemuth, M. (2015). Wertschöpfung von Molkereien. MEG Milch Board, Göttingen.

Key, N., Runsten, D. (1999). Contract farming, smallholders, and rural development in Latin America: the organization of agroprocessing firms and the scale of outgrower production. *World Development* 27: 381–401.

Khan, M.H. (2001). Agricultural taxation in developing countries: a survey of issues and policy. *Agricultural Economics* 24: 315–328.

Kreß, B. (2005). Die EU-Agrarreform vom 26.6.2003 und die Konsequenzen für die Milchwirtschaft – Expertenbefragung in Dänemark, Österreich und Deutschland. Dissertation, Technical University of Munich.

Koester, U., Cramon-Taubadel, S. (2019). Besonderheiten der landwirtschaftlichen Kreditmärkte. Leibnitz Institute of Agricultural Development in Transition Economics, IAMO Discussion Papers, No. 185, Halle.

Lassen, B., Isermeyer, F., Friedrich, C. (2009). Regional changes in German dairy production. *Agrarwirtschaft* 58, booklet 5/6: 238–247.

Li, L. (2007). Multinationality and performance: A synthetic review and research agenda. *International Journal of Management Reviews* 9(2): 117–139.

Mehlhose, C., Hunecke, C., Spiller, A., Brümmer, B. (2019). Der Markt für Milch- und Milcherzeugnisse im Jahr 2018. *German Journal of Agricultural Economics* 68: 62–84.

Meyer, J., Theuvsen, L. (2017). Intensive dairy farming in Northern Germany: development and impact of the New Fertilizer Act. In: Agrarian Perspectives XXVI.

Competitiveness of European Agriculture and Food Sectors, Proceedings of the 26th International Conference, 13–15 September 2017, Prague: 219–225.

Meyer, J., Feil, J.H., Schaper, C. (2019). Internationalization strategies in the German dairy industry and their influence on the economic performance of firms. *International Journal of Food System Dynamics* 10(4): 332–346.

MIV (2017). Wohin die Milch fließt. German Association of Dairy Industry, Berlin.

OECD (2001). Competition Issues in Joint Ventures. OECD Policy Roundtables, OECD, Paris.

OECD, FAO (2015). OECD-FAO Agricultural Outlook 2015–2024. OECD Publishing., Paris.

OECD, FAO (2019). OECD-FAO Agricultural Outlook 2019–2028. OECD Publishing, Paris.

Perlmutter, H.V. (1969). The Tortuous Evolution of the Multinational Corporation. *Columbia Journal of World Business* 4: 9–18.

Porter, M.E. (1990). The competitive advantage of nations. *Harvard Business Review* March-April 1990: 73–93.

Qian, G. (2002). Multinationality, product diversification and profitability of emerging US small- and medium-sized enterprises. *Journal of Business Venturing* 17: 611–633.

Reardon, T., Barrett, C.B. (2000). Agroindustrialization, globalization, and international development: An overview of issues, patterns, and determinants. *Agricultural Economics* 23: 195–205.

Reardon, T., Barrett, C.B., Berdegué, J.A., Swinnen, J.F.M. (2009). Agrifood Industry Transformation and Small Farmers in Developing Countries. *World Development* 37: 1717–1727.

Riley, S. (2017). Milking the Latin American dairy market. Dairy reporter.com, available at: https://www.dairyreporter.com/Article/2017/09/21/Milking-the-Latin-American-dairy-market (accessed 20 August 2019).

Theuvsen, L., Ebneth, O. (2005). Internationalization of Cooperatives in the Agribusiness: Concepts of Measurement and their Application. In: Theurl, T. (Ed.), *Strategies for Corporation*, Shaker, Aachen: 395–419.

Schlippenbach, V., Pavel, F. (2011). Konzentration im Lebensmitteleinzelhandel: Hersteller sitzen am kürzeren Hebel. *DIW Wochenbericht* 78(13): 2–9.

Stiglitz, J.E. (1987). Some Theoretical Aspects of Agricultural Policies. *The World Bank Observer* 2(1): 43–60.

Sutton, J. (2003). Understanding the rise in global concentration in the agri-food sector: a background paper. In: *OECD Conference on Changing Dimensions of the Food Economy*: 6–7.

Top agrar (2019). Neue Düngeverordnung soll ab Mai 2020 gelten. Top agrar online, available at: https://www.topagrar.com/acker/news/neue-duengeverordnung-soll-ab-mai-2020-gelten-10289806.html (accessed 20 September 2019).

Trademap (2019). Trade data of groups 0401-0406. Trademap.org, available at: https://www.trademap.org/Index.aspx (accessed 20 June 2019).

Tybout, J.R. (2000). Manufacturing firms in developing countries: How well do they do, and why? *Journal of Economic Literature* 38,(1): 11–44.

Verbeek, M. (2004). *A Guide to Modern Econometrics*. John Wiley & Sons Ltd., Hoboken, NJ.

Wilkinson, J. (2004). The food processing industry, globalization and developing countries. *Electronic Journal of Agriculture Development Economics* 1: 184–201.

Wind, Y., Douglas, S.P., Perlmutter, H.V. (1973). Guideline for Developing International Strategies. *Journal of Marketing* 37(2): 14–23.

ZMB (2016). ZMB Jahrbuch Milch 2016. ZMB Zentrale Milchmarkt Berichterstattung GmbH, Berlin.

Appendix

Table 1. Descriptive statisitcs

Variable		Mean	Std. Dev.	Min	Max	Observations
ID	overall	9.54	5.28	1	18	N = 138
	between		5.34	1	18	n = 18
	within		0	9.54	9.54	T-bar = 7.67
Year	overall	4.57	2.27	1	8	N = 138
	between		0.44	4	6	n = 18
	within		2.24	1.07	8.07	T-bar = 7.67
EBIT-margin (%)	overall	2.33	1.98	-5.2	8.7	N = 138
	between		1.50	0.2	4.8625	n = 18
	within		1.323685	-3.70	6.28	T-bar = 7.67
Internationalization-Strategy (dummy)	overall	0.29	0.46	0	1	N = 138
	between		0.46	0	1	n = 18
	within		0.00	0.29	0.29	T-bar = 7.67
Organization (dummy)	overall	0.46	0.50	0	1	N = 138
	between		0.51	0	1	n = 18
	within		0	0.46	0.46	T-bar = 7.67
Turnover (million Euros)	overall	1927.85	3107.64	132.08	12110	N = 138
	between		3120.33	183.58	10756.13	n = 18
	within		429.95	-628.48	3407.93	T-bar = 7.67
Gearing Ratio (%)	overall	229.82	180.05	47.1	1790.80	N = 137
	between		141.32	50.85	657.27	n = 18
	within		117.64	-82.25	1363.35	T-bar = 7.67
Dairy-Price-Index (index value)	overall	201.18	30.24	153.77	242.75	N = 138
	between		1.69	196.13	201.60	n = 18
	within		30.21	153.35	247.80	T-bar = 7.67
Frischmilch (dummy)	overall	0.52	0.50	0	1	N = 138
	between		0.50	0	1	n = 18
	within		0.12	0.15	1.15	T-bar = 7.67
Frischmilchprodukte (dummy)	overall	0.75	0.44	0	1	N = 138
	between		0.43	0	1	n = 18
	within		0.12	0.12	1.12	T-bar = 7.67
Käse (dummy)	overall	0.88	0.32	0	1	N = 138
	between		0.32	0	1	n = 18
	within		0	0.88	0.88	T-bar = 7.67
Trockenprodukte (dummy)	overall	0.73	0.44	0	1	N = 138
	between		0.46	0	1	n = 18
	within		0.00	0.73	0.73	T-bar = 7.67
Butter (dummy)	overall	0.59	0.49	0	1	N = 138
	between		0.50	0	1	n = 18
	within		0.00	0.59	0.59	T-bar = 7.67

Source: Authors' own depiction based on Meyer *et al.*, 2019.

VII Summary of Main Findings, General Conclusions and Future Research

7.1 Summary of Main Findings

The German dairy sector has experienced increasing liberalization in recent years. On the one hand, this has opened up new opportunities and possibilities but on the other hand, the dairy industry is also facing increasing international competition for market share and heavy price pressure on global markets. At the same time, domestic demands on milk production are increasing, for instance with regard to environmental requirements and husbandry conditions. This dissertation analyzes the development of milk production and the impact of the New Fertilizer Act on it (Chapter II) together with drivers of the development towards large dairy herds in Germany (Chapter III). With regard to the dairy industry itself, this dissertation examines how competitiveness has developed on the international market for milk and dairy products (Chapter IV) and how different internationalization strategies affect corporate success (Chapter V). Based on these results, strategic lines of development for the German dairy industry are shown, as well as ways for their implementation on the basis of a market entry in Brazil (Chapter VI). The results of the contributions are summarized below.

The results in Chapter II show the enormous importance of intensive dairy regions with a milk production of more than 2.000kg milk/ha. In 2015, these regions accounted for 59.8% of German milk production. These regions are even more important in terms of growth in milk production. Intensive dairy regions already accounted for 76.2% of milk production growth in 2010. In 2015, these regions accounted for 78.5% of volume growth. Over the entire observation period from 2010 to 2015, their share was 69.9%. These figures impressively demonstrate the importance of these regions for milk production in Germany. In the two largest producer countries Bavaria and Lower Saxony, the intensive regions contributed significantly to milk production over the period under review with 75.9% and 77.8% respectively. The high correlation between milk volume at district level and grassland area of 0.889 underlines the importance of grassland for milk production. Its share of total farmland is up to 49.2% in high-intensity regions with a milk production of more than 3,000 kg. This result also underlines the results from the literature which show that grassland locations are favorable locations for milk production due to their low opportunity costs (Lassen *et al.*, 2009). Due to their high intensity, the intensive regions show a high accumulation of liquid manure (Chamber of Agriculture, 2017). Against this background, the Fertilizer Act will affect the farms in these regions in particular due to the strict limit of 170 kg total nitrogen of animal origin/ha and the abolition of the derogation option which previously allowed 230 kg N/ha from animal origin to be spread on grassland. In particular, the increase in demand for land solely for spreading liquid manure and the export of nutrients resulting from the surpluses are associated with rising production costs. Cost increases from past increases in the cost of arable land and grassland amount to 0.7Ct./kg Energy Corrected Milk (ECM) The additional disposal costs for the farms analyzed would amount to additional costs of 0.4Ct./kg ECM (Niemann, 2016).

Chapter III analyses the development of large dairy herds in Germany and the factors influencing it in the period from 1992 to 2018. The results show that only herds with more than 100 cows grow continuously over the entire period, both in terms of the number of cows and the number of farms. Thereby milk quota has clearly had a negative effect on the development of the large herds. By contrast, the sharp drop in interest rates in the period under review and thus the declined capital costs had the most positive impact on the development of large herds. The milk price and business expectations also have a significant influence on the development of the large herds. Our results thus essentially confirm those of the literature. However, we did not find any significant influence of opportunity costs on the development of large herds. The reason for this can be explained by the strong expansion of dairy farming in regions dominated by grassland, which are characterized by low opportunity costs due to the limited possibilities for using grassland (Lassen *et al.*, 2009).

The expiry of the milk quota at the end of 2015 paved the way for further growth in milk production and thus for the further growth of farms and herd sizes. However, this is likely to be slowed by several factors in the future. Rising production costs, e.g. due to increased environmental regulations and construction costs are leading ceteris paribus to a decline in profitability of farms. The strong increase in agricultural loans leads to higher debt services which results with ongoing volatile prices to a higher liquidity risk of farms. On the one hand, this should reduce the demand for loans and, on the other hand, make access to capital more difficult. In addition, imperfect information is increasing, for example due to price volatility, but also due to uncertain prospects, e.g.

with regard to husbandry conditions which might also lead to a decline in investments and thus in growth of large herds.

Overall, the growth of the large herds will continue due to their advantages in terms of production costs, but is likely to slow down despite the abolition of the quota due to these aspects. Future investments and thus growth will concentrate even more on the most economically efficient farms.

Chapter IV analyzes the international competitiveness of the German dairy industry on the basis of foreign trade data. The results show that there is no "international competitiveness". Rather, the international competitiveness of the countries under consideration, expressed in the Relative Trade Advantage (RTA) according to Vollrath (1991), depends on the product group under consideration. Exceptions to this are New Zealand and Belarus, which are highly competitive across all product groups considered, which is also confirmed by the literature (IFCN, 2013; Fahlbusch, 2014). Germany is competitive for the product groups buttermilk, cream and yoghurt, whey and concentrated milk products. The negative RTA values for non-concentrated dairy products, butter and cheese, indicate competitive disadvantages, with the competitiveness of butter and cheese improving over time. Across all product groups, competitiveness increased marginally over time. However, unlike most competitors, such as New Zealand and the USA, this increase is not due to an increase in Relative Export Advantage (RXA), but to an increase in Relative Import Advantage (RMP). The reasons for this include the development of market shares. These have increased amongst competitors over time, while Germany's market share has declined across almost all product groups and also in total. The reason for this is that although Germany's exports have risen, those of the competitors and the overall market have risen faster.

With regard to the condition of gaining or at least retaining market shares within the definition of competitiveness, this development puts the German dairy industry at a competitive disadvantage vis-à-vis its competitors. Germany, like other European countries, was unable to expand its milk production until 2007 due to the CAP and in particular the milk quota, which is one reason for the developments observed (Kreß, 2005; Kirner, 2009). However, the results show that even the increased quota volumes resulting from the abolition of the milk quota and the resulting increase in milk production in Germany were not sufficient to win back or retain shares in the export market. Against the background of requirements increasing further, especially with regard to fertilization legislation, further growth is expected to slow, particularly in the intensive regions that make such a decisive contribution to growth to date. It is unlikely that this picture will change despite the abolition of the milk quota. In addition, the evaluation of production costs also shows cost advantages in milk production for important competitors (IFCN 2013). The high absolute values for market shares indicate a highly competitive situation in the past. The results show that competitors have caught up and Germany's relative competitive position has deteriorated.

In addition, the results also show different export orientation with regard to the product groups investigated in the countries considered; also the importance of trade agreements, both bilateral and multilateral, which are reflected in the trading partners of the countries in question. The export of cheese and curd accounts for the largest share of German exports at 45.3% in 2017, while by far the most important customers not only for Germany but also for the European countries in question, are the countries within the single European market. In Germany, their share of total exports declined in all product groups considered over the period under review but at 83.3% they were still at a high level. The most important customer for Germany outside the European Union is China with a 3% share of total exports.

Chapter V analyzed the influence of different internationalization strategies on the economic success of 18 dairies, of which 16 are headquartered in Germany. The German dairy companies considered in this study represent 60% of the total turnover of the German dairy industry and 90.4% of the German dairy companies' foreign turnover. Based on the extended EPRG model of Perlmutter (1969) by Wind *et al.* (1973), three internationalization strategies were assigned to the companies investigated: the international, the global and the multinational strategy. Over the period considered from 2010 to 2017, the results of the study show a significant positive effect of the multinational strategy on the corporate success of the companies examined. A product portfolio with cheese and dried dairy products as well as the gearing ratio, also had a positive effect on the economic success of the companies, measured as EBIT margin. The latter, however, has only a marginal effect. Butter, on the other hand, had a negative impact on the company's success in the product portfolio.

The reasons for the strong positive effect of the multinational strategy on the company's success, based on the extended EPRG model, indicate that the competitive advantages through localization and differentiation outweigh those from globalization and standardization. This enables companies to better adapt to local demand conditions and build brands that generate higher sales. In addition, due long transport distance limits due to shelf life, especially for fresh milk products such as yoghurt, milk drinks and fresh cheese, they are able to offer a different product portfolio on distant markets and thus better exploit the market potential. This result is also supported by the literature, which also confirms a positive influence of multi-nationality and diversification on business success (Zou and Stan, 1998; Harzing, 2000; Qian, 2002; Ebneth, 2006; Sousa *et al.*, 2008). Due to the high factor demand of the multinational strategy, but also higher sunk costs compared to the international strategy in particular, the multinational strategy is implemented by the more successful companies.

The legal form of the companies in question has no significant influence on the success of the company. Therefore, we cannot confirm that being a cooperative has a negative impact on the success of the company and thus does not confirm the results of other studies (Anderson and Henehan, 2005; Ebneth, 2006). At the very least, this applies to the companies surveyed. However, with regard to the corporate forms of the companies, it can be seen that the multinational strategy is implemented exclusively by private companies with their headquarters in Germany. Apart from one cooperative pursuing the global strategy, all German cooperative dairies pursue the international strategy.

Based on the previous results, Chapter VI examines the opportunities and risks for German dairies entering the Brazilian dairy sector, the local hurdles and difficulties and how companies can react to them. The Brazilian market offers enormous potential for cost-effective milk production in view of the almost abundant natural resources and the opportunity to participate in the Latin American market, one of the fastest growing in the world (Riley, 2017). Also, due to the significantly lower concentration in the Brazilian food retail trade, the dairy companies are also likely to have a much better negotiating position with customers than in Germany. This should in turn have a positive effect on price conditions and the listing of products (Farina, 2002; Schlippenbach and Paval, 2011; Abras, 2018).

Risks and hurdles lie both in agricultural production and in the dairy industry itself. With regard to Brazilian agriculture, four main aspects can be identified that present potential companies with challenges. These are the low level of professionalization of farmers, quality and sanitary aspects, the drafting and enforcement of contracts between farmers and dairies, and infrastructure in the form of poor roads and, in some cases, long distances with small collection volumes. In the Brazilian dairy industry itself, four major weaknesses could also be identified: low professionalization of human resources, low transmission of technology, costly overcapacities and low investment in research, technology and innovation, which many managers see more as superfluous expenditure than as investment.

With regard to any possible market entry, the already existing overcapacities mean that the form of entry should rather be a joint venture rather than new construction and thus the creation of further capacities. The number of potential companies for such a possible entry is limited, at least in the short term, by the high factor requirements, e.g. for capital and management, in addition with regard to governance structures, especially with regard to cooperatives (Theuvsen and Ebneth, 2005; Brokelmann and Gäde, 2017). For the time being, any entry into the Brazilian market, at least in the short term, is likely to be reserved for some German private dairies due to governance structures and economic efficiency.

In contrast to an expansion of production or a joint venture, for example in Germany, companies that are thinking of entering the Brazilian market must also focus on milk production in addition to actual milk processing. This is an essential aspect that applies in many ways. At least for the time being, companies would have to develop their own appropriate requirements and create incentive systems for farmers to improve quality and comply with them. possible way to implement this would be contract farming. This could create an incentive for farmers, for example, to make necessary investments and at the same time bind them to the company as suppliers. Farmers could also receive direct financial support from the companies with which they conclude contracts (Koester and Cramon-Taubadel, 2019).

The companies also play a central role with regard to training and counselling opportunities, both for their own employees and for farmers, since despite the progress made, such offers are not as prominent as in Germany. Low levels of professionalization are

also a major reason for the low transmission of technology. For this reason, companies should also provide advice on farm, herd and feed management. On the one hand this has an influence on quality parameters, on the other hand production costs can be reduced. The cultural proximity and the fact, despite a decreasing tendency, that many people still speak German in the GFM region would make this aspect particularly attractive to German companies.

7.2 General Conclusions and Future Research

The German dairy industry is one of the most important sectors in the German agribusiness and will remain so in the future. In the future, however, the dairy industry and German milk producers will face the great challenge of meeting increasing demands from the domestic market, for example over environmental requirements and societal demands regarding GMO-free feeding and husbandry conditions, while competition on the international market and price pressure continue to increase. Although new trends, such as regional and organic products offer dairy companies opportunities to bring new products with higher added value onto the domestic market, this is taking place against the backdrop of a saturated domestic market. On the other hand, a look at these markets shows that, for example, the market for organic milk is growing rapidly but its importance is still low and the market for conventional milk will still be of enormous importance for dairies. For example, the share of organic milk in the total milk delivered was 3.7% in 2018 (Mehlhose *et al.*, 2019, BLE, 2019). The fact that almost half of the milk delivered is exported is also unlikely to change in the foreseeable future (MIV, 2019). Nor can any negotiating position vis-à-vis the highly concentrated retail trade be expected to improve in the future, so that here too, no positive impulses for an improvement can be expected in the on average weak economic situation of German dairy companies (Brokelmann and Gäde, 2017).

Internationalization will therefore remain a major driver in the industry. This work has shown that the form of internationalization is an essential factor in the economic success of companies (Ebneth, 2006; Meyer *et al.*, 2019). Due to the domestic market's increasing demands, which are reflected in particular in milk production rising costs, and the already difficult situation with regard to good added value in the export of less differentiated products, this situation will present dairies with major challenges, in particular those which have hitherto relied on the simple export of milk products (Jürgens, *et al.*,

2015; Brokelmann and Gäde, 2017). Under these circumstances the relative competitiveness of the German dairy sector vis-à-vis competitors on the world market will decrease, as price is the most sought-after product feature on international markets (BVE, 2017).

German dairy companies should therefore examine in future whether, and in what form, they could enter other countries' dairy markets in order to develop promising markets in future, not only through exports but also through locally produced products. The work shows that the advantages that companies gain from differentiation and localization exceed that of standardization and globalization. This advantage is likely to increase even further with the expected demand and production growth in Asia, as well as stronger increase in demand for fresh milk products (OECD and FAO, 2019). Existing activities can and should be continued and potential on the domestic market pursued. In addition, companies should make use of their know-how by deploying it to their advantage in other countries; thereby, taking the attributes of quality and food safety associated with German food as competitive advantages for their image and brand building abroad (Brokelmann and Gäde, 2017; BVE, 2017). However, internationalization in the form of direct investment places considerable demands on any company's resources, which in turn requires sufficient capital resources and financial strength to implement such a strategy. This is likely to pose challenges to many cooperative dairies in particular. Furthermore, hurdles regarding economic and political framework conditions in many countries are also high. This can be seen both in the literature and in this paper, using Brazil as an example (Grant and Nippa, 2006; Theuvsen *et al.*, 2010; Pfeiffer, 2015; Brokelmann and Gäde, 2017).

However, both the results and the success of the companies show that their efforts can prove worthwhile. Referring back to the opening quotation and the current examples, both from German and foreign dairies, companies are increasingly pursuing this form of internationalization, such as Ehrmann in Brazil and Arla, which is currently investing in Nigeria. More examples arise such as in the meat industry, where the company Tönnies is starting to set up a slaughterhouse in China (Handelsblatt, 2019; Cornall, 2019). In addition to economic advantages for those companies, entry into production in other countries offers the advantage that these countries are able to participate in the global economy, both economically and socially, through positive employment effects

(Pfeiffer, 2015). Private German dairies have already embarked on this path. Cooperative dairies will face greater challenges in this form of internationalization.

Future research should on the one hand aim at identifying potential markets worldwide and analyzing them with regard to local conditions and the resulting requirements for dairy companies. With regard to cooperative dairies, it should be examined whether and under which conditions their members would support a possible internationalization through direct investment abroad. On the other hand, expert interviews could be conducted with executives of cooperative dairies that are already pursuing this strategy. In this way particular possibilities arising for German cooperative dairies could be analyzed with regard to entering new markets by means of direct investment and how this strategy could be implemented in concrete terms. In future, this research could best be carried out in cooperation with companies or associations from the dairy industry. In a very recent study, for example, the German Dairy Industry Association points out not only the need for research into product and process technologies but also the need for research into future business models as a result of the milk surplus in developed countries (Block, 2019). Such cooperation could contribute additional expertise and further improve the results.

References

ABRAS (2018). Economia e Pesquisa » Ranking Abras » As 500 maiores – ABRAS. ABRAS – Associação Brasileira de Supermercados, available at: http://www.abrasnet.com.br/economia-e-pesquisa/ranking-abras/as-500-maiores/ (accessed 23 September 2019).

Anderson, B.L., Henehan, B. (2005). What gives agricultural co-operatives a bad name? *International Journal of Cooperative Management*, 2(2): 9–14.

BLE (2019). Kuhmilchlieferungen der Erzeuger an deutsche milchwirtschaftliche Unternehmen. Federal Office for Agriculture and Food, Bonn.

Block, H.J. (2019). Eine leistungsfähige Forschung für die Milchwirtschaft in Deutschland. Study of the German Dairy Association. German Dairy Associaton, Berlin.

Brokelmann, V., Gäde, S. (2017). Branchenstudie Molkereiwirtschaft 2017. HSH Nordbank AG.

BVE (2017). Fakt ist: Lebensmittelexport. Fakt ist No. 6, Fedration of German Food and Drink Industries, Berlin.

Chamber of Agriculture (2017). Nährstoffbericht in Bezug auf Wirtschaftsdünger für Niedersachsen 2015/2016. Chamber of Agriculture of Lower Saxony.

Cornall, J. (2019). Arla commits to sustainable dairy sector in Nigeria. Dairy reporter.com, available at: https://www.dairyreporter.com/Article/2019/09/16/Arla-commits-to-sustainable-dairy-sector-in-Nigeria?utm_source=newsletter_daily&utm_medium=email&utm_campaign=16-Sep-2019&c=oVcyMhVABWGgxJntPJsv2qOnjlTKzGn0&p2 (accessed 29 September 2019).

Ebneth, O.J. (2006). Internationalisierung und Unternehmenserfolg, ein Vergleich europäischer Molkereigenossenschaften. *Schriften der Gesellschaft für Wirtschafts- und Sozialwissenschaften des Landbaus e.V.* 41: 363–374.

Fahlbusch, M. (2014). Price Formation and the Measurement of Market Power on the International Dairy Markets. PhD thesis at the chair of Agricultural Market Analysis, University of Göttingen.

Farina, E.M.M.Q. (2002). Consolidation, multinationalisation, and competition in Brazil: impacts on horticulture and dairy products systems. *Development Policy Review* 20: 441–457.

Grant, R.M., Nippa, M. (2006). *Strategisches Management*. Pearson Studium, Hallbergmoos.

Handelsblatt 2019. Tönnies investiert mit Partner 500 Millionen Euro in China. Online platform of Handelsblatt, available at: https://www.handelsblatt.com/unternehmen/mittelstand/familienunternehmer/schlachtbetrieb-toennies-investiert-mit-partner-500-millionen-euro-in-china/25061860.html (accessed 29 September 2019).

Harzing, A.W. (2000). An empirical analysis and extension of the Bartlett and Ghoshal typology of multinational companies. *Journal of International Business Studies* 31(1): 101–120.

IFCN (2013). Dairy Report 2013. International Farm Comparison Network.

Jürgens, K., Fink-Keßler, A., Poppinga, O., Wohlgemuth, M. (2015). Wertschöpfung von Molkereien. MEG Milch Board, Göttingen.

Kirner, L. (2009). Auswirkung der vollständigen Implementierung des Health-Check auf landwirtschaftliche Betriebe. *Ländlicher Raum*, online journal of the Austrian ministry for Agriculture, Forestry, Environment and Water Conservancy, available at: https://www.bmnt.gv.at/land/laendl_entwicklung/zukunftsraum_land_masterplan/Online-Fachzeitschrift-Laendlicher-Raum/archiv/2009/kirner.html (accessed 20 August 2019).

Koester, U., Cramon-Taubadel, S. (2019). Besonderheiten der landwirtschaftlichen Kreditmärkte. Leibnitz Institute of Agricultural Development in Transition Economics, IAMO Discussion Papers, No. 185, Halle.

Kreß, B. (2005). Die EU-Agrarreform vom 26.6.2003 und die Konsequenzen für die Milchwirtschaft – Expertenbefragung in Dänemark, Österreich und Deutschland. Dissertation, Technical University of Munich.

Lassen, B., Isermeyer, F., Friedrich, C. (2009). Regional changes in German dairy production. *Agrarwirtschaft* 58, booklet 5/6: 238–247.

Mehlhose, C., Hunecke, C., Spiller, A., Brümmer, B. (2019). Der Markt für Milch- und Milcherzeugnisse im Jahr 2018. *German Journal of Agricultural Economics* 68: 62–84.

Meyer, J., Feil, J.H., Schaper, C. (2019). Internationalization strategies in the German dairy industry and their influence on the economic performance of firms. *International Journal of Food System Dynamics* 10(4): 332–346.

MIV (2019a). Wohin die Milch fließt. Dairy Industry Association of Germany, available at:

https://milchindustrie.de/wp-content/uploads/2018/11/Wohin-die-Milch-flie%C3%9Ft-2018.pdf (accessed 22 September 2019).

Niemann, F. (2016). Novellierung der Düngeverordnung: Anpassungsstrategien von Milchviehbetrieben in einer Intensivregion. Master Thesis, Georg-Ausgust-University Göttingen, Göttingen.

OECD, FAO (2016). OECD-FAO Agricultural Outlook 2016–2025. Paris, OECD Publishing,

Perlmutter, H.V. (1969). The Tortuous Evolution of the Multinational Corporation. *Columbia Journal of World Business* 4: 9–18.

Pfeiffer, B. (2015). “Upgrading” in Wertschöpfungsketten – Global und Regional? German Institute of Global Studies, Leibnitz Institut für Globale und Regionale Studien, GIGA Focus No. 8.

Qian, G. (2002). Multinationality, product diversification and profitability of emerging US small- and medium-sized enterprises. *Journal of Business Venturing* 17: 611–633.

Riley, S. (2017). Milking the Latin American dairy market. Dairy reporter.com, available at:
https://www.dairyreporter.com/Article/2017/09/21/Milking-the-Latin-American-dairy-market (accessed 20 August 2019).

Schlippenbach, V., Pavel, F. (2011). Konzentration im Lebensmitteleinzelhandel: Hersteller sitzen am kürzeren Hebel. *DIW Wochenbericht* 78(13): 2–9.

Sousa, C.M.P., Martinez-López, F.J., Coelho, F. (2008). The determinants of export performance: A review of research in the literature between 1998 and 2005. *International Journal of Management Reviews* 10(4): 343–374.

Theuvsen, L., Ebneth, O. (2005). Internationalization of Cooperatives in the Agribusiness: Concepts of Measurement and their Application. In: Theurl, T. (Ed.), *Strategies for Corporation*, Shaker, Aachen: 395–419.

Theuvsen, L., Janze, C., Heyder, M. (2010). Agribusiness in Germany 2010. On the Way to New Markets, Ernst & Young, Stuttgart.

Vollrath, T.L. (1991). A Theoretical Evaluation of Alternative Trade Intensity Measures of Revealed Comparative Advantage. *Weltwirtschaftliches Archiv* 127(2): 265–280.

Wind, Y., Douglas, S.P., Perlmutter, H.V. (1973). Guideline for Developing International Strategies. *Journal of Marketing* 37(2): 14–23.

Zou, S., Stan, S. (1998). The determinants of export performance: A review of the empirical literature between 1987 and 1997. *International Marketing Review* 15(5): 333–356.

Curriculum Vitae

Johannes Meyer

Personal Information

Date of birth	14.04.1988
Place of birth	Nienburg/Weser
Address	Kroger Straße 5, 31608 Marklohe
E-Mail/Telephone	johannes.meyer@agr.uni-goettingen.de / 01701834924

Academic Career

11/2015 – today	**Research associate** (PhD Candidate), University of Göttingen, chair of Agribusiness Management
707/2012 – 06/2015	**Student Assistant**, University of Göttingen, chair of Agribusiness Management
10/2012 – 03/2015	**Master's degree in Agricultural Science** Specializing in WiSoLa, University of Göttingen. Topic of Master Thesis: "Validation of a key figure-based concept for measuring risk-bearing capacity of farms with the help of the VR Rating Agrar"
10/2010 – 10/2013	**Bachelor's degree in Agricultural Science** Specializing in WiSoLa, University of Göttingen. Topic of Bachelor Thesis: "Development of a program for rating preparation and support for the implementation of the Improvement of farmers' ratings"
08/2007 – 07/2009	**Training as a farmer**

Internships

04/2014 – 06/2014	**Internship** Ländliche Betriebsgründungs- und Beratungsgesellschaft mbH, Göttinngen
04/2012 – 06/2012	**Internship** WGZ Bank AG, Münster

Languages	**Deutsch** (mother tongue) **English** (Business fluent, C1 level)
IT Knowledge	**MS-Office, Stata** (Good knowledge)

Declaration of Own Contribution to Work Performed

Hereby I declare the share of the contributions made in the dissertation.

In the first contribution in Chapter II entitled "Intensive Dairy Farming in Northern Germany: Development and Impact of the New Fertilizer Act", which was written in cooperation with Prof. Dr. Ludwig Theuvsen, the following areas were taken over by me: Idea and conception of the study with consulting by Prof. Dr. Ludwig Theuvsen, execution of the calculation and interpretation of the results with consulting by Prof. Dr. Ludwig Theuvsen. In addition, the writing and revision of the paper was done with the advice Prof. Dr. Ludwig Theuvsen.

In the second contribution in Chapter III entitled "Drivers of Large Herd Sizes in German Dairy Farming: Development and Outlook", which was written in cooperation with Prof. Dr. Jan-Henning-Feil and Dr. Christian Schaper, the following areas have been taken over by me: Idea and conception of the study with consulting by Prof. Dr. Jan-Henning Feil and Dr. Christian Schaper, execution of the calculation and interpretation of the results with consulting by Dr. Sebastian Lakner, Prof. Dr. Jan-Henning-Feil and Dr. Christian Schaper.

In the third contribution in Chapter IV entitled "Assessing the International Competitiveness of the German Dairy Industry by Analysing Foreign Trade", which was written in cooperation with Dr. Sebastian Lakner, Prof. Dr. Jan-Henning-Feil and Dr. Christian Schaper, the following areas have been taken over by me: Idea and conception of the study with consulting by Dr. Sebastian Lakner, Prof. Dr. Jan-Henning Feil and Dr. Christian Schaper, execution of the calculation and interpretation of the results with consulting by Dr. Sebastian Lakner, Prof. Dr. Jan-Henning-Feil and Dr. Christian Schaper. In addition, the writing and revision of the paper was done with the advice of Dr. Sebastian Lakner, Prof. Dr. Jan-Henning-Feil and Dr. Christian Schaper.

In the fourth contribution in Chapter V entitled "Internationalization Strategies in the German Dairy Industry and their Influence on the Economic Performance of Firms", which was written in collaboration with Prof. Dr. Jan-Henning-Feil and Dr. Christian Schaper, I took over the following areas: Idea and conception of the study with consulting by Prof. Dr. Jan-Henning Feil and Dr. Christian Schaper, execution of the calculation and interpretation of the results with consulting by Prof. Dr. Jan-Henning-Feil and

Dr. Christian Schaper. In addition, the writing and revision of the contribution took place with the advice of Prof. Dr. Jan-Henning-Feil and Dr. Christian Schaper.

In the fifth contribution in Chapter VI entitled "German Brazilian dairy supply chain integration: Prospects and ways", which was written in collaboration with Dr. Caetano Beber, I took over the following areas: Idea and conception of the study in collaboration with Dr. Caetano Beber, execution of the calculation and interpretation of the results in collaboration with Dr. Caetano Beber.

www.ingramcontent.com/pod-product-compliance
Ingram Content Group UK Ltd.
Pitfield, Milton Keynes, MK11 3LW, UK
UKHW021652190726
13853UKWH00001B/220

9 783736 972124